本科专业英语系列教材

**教材出版获贵州大学
安全工程省级一流
专业建设项目资助**

ENGLISH FOR SAFETY ENGINEERING

安全工程专业英语

主编 袁 梅 江泽标 李波波

重庆大学出版社

图书在版编目(CIP)数据

安全工程专业英语 / 袁梅, 江泽标, 李波波主编. -- 重庆: 重庆大学出版社, 2022.8
本科专业英语系列教材
ISBN 978-7-5689-3304-9

Ⅰ.①安… Ⅱ.①袁… ②江… ③李… Ⅲ.①安全工程－英语－高等学校－教材 Ⅳ.①X93

中国版本图书馆CIP数据核字(2022)第078955号

安全工程专业英语

主编 袁 梅 江泽标 李波波
责任编辑：牟 妮　　版式设计：牟 妮
责任校对：关德强　　责任印制：赵 晟

*

重庆大学出版社出版发行
出版人：饶帮华
社址：重庆市沙坪坝区大学城西路21号
邮编：401331
电话：(023)88617190　88617185（中小学）
传真：(023)88617186　88617166
网址：http://www.cqup.com.cn
邮箱：fxk@cqup.com.cn（营销中心）
全国新华书店经销
重庆新生代彩印技术有限公司

*

开本：787mm×1092mm　1/16　印张：8.75　字数：243千
2022年8月第1版　2022年8月第1次印刷
ISBN 978-7-5689-3304-9　定价：49.00元

本书如有印刷、装订等质量问题，本社负责调换
版权所有，请勿擅自翻印和用本书
制作各类出版物及配套用书，违者必究

序

安全是人类的本能要求。在我国,"安全"一词早在汉代就已出现。在现代汉语中,"安全"作为一个词语,在各种辞书中都有着"没有危险,不受威胁"这一基本相同的解释。2000多年来,中国人正是以安心、安身为基本人生观,一直以居安思危的态度促其实现。在人类社会发展进程中,著名的马斯洛层次需求论,也将"安全"作为人类需求的五个等级之一纳入层次需求理论的金字塔模型,并使其成为引发动机和激励的力量。

当前,是我国全面建成小康社会、实现第一个百年奋斗目标之后,乘势而上开启全面建设社会主义现代化国家新征程、向第二个百年奋斗目标进军的关键时期,发展中的矛盾和问题不仅集中体现在高质量发展上,而且也体现在安全发展上,习近平总书记强调保证国家安全是头等大事,提出总体国家安全观,涵盖政治、军事、国土、经济、文化、社会、科技、网络、生态、资源、核、海外利益、太空、深海、极地、生物等诸多领域,要求全党全国人民增强斗争精神、提高斗争本领,落实防范化解各种风险的领导责任和工作责任。

毫无疑问,安全发展贯穿于国家发展各领域全过程,注重防范化解影响我国现代化进程的重大风险是我党的一贯要求,国泰民安是人民群众最基本、最普遍的愿望。在社会主义经济建设中,安全是发展的前提,发展是安全的保障,没有发展就没有安全的物质基础,没有安全就没有发展的基本支撑。奋进新时代、开启新征程,需要我们把发展质量与安全的关系摆在更加突出的位置。

习近平总书记在清华大学考察时强调:"我国高等教育要立足中华民族伟大复兴战略全局和世界百年未有之大变局,心怀'国之大者',把握大势,敢于担当,善于作为,为服务国家富强、民族复兴、人民幸福贡献力量"。作为普通高等学校本科专业之一的安全工程专业,为适应国家社会经济发展需要,培养了大批掌握安全科学、安全技术、安全管理及职业健康基本理论、基础知识和基本技能,具备专门从事安全工程设计、研究、检测、评价、监察及管理等工作能力的高素质复合型工程技术专业人才。

贵州大学开设的安全工程专业英语课程，主要是针对矿山安全领域复杂工程问题，培养学生与国际同行及社会公众进行有效沟通和交流，使之具备一定的国际视野和跨文化交流与合作的能力，契合国家经济发展和学生就业的现实需要。袁梅教授、江泽标副教授、李波波教授借助多年的教学经验，按教学大纲自编的《安全工程专业英语》教材在贵州大学安全工程本科专业使用多年，受到广大师生的好评，这次又在自编教材的基础上结合安全工程专业现状和未来的发展动向进行遴选和整合，编写了这本教材。

本教材结合安全工程专业的培养目标，编写了18个单元，教材中所有单元均节选自国内已出版的安全工程专业英语教材、相关教辅和网上公开相关资料。其中涉及煤矿安全的内容重点以贵州煤矿"第一杀手"的瓦斯灾害为重点，兼顾矿井通风及矿井火灾等；非煤矿安全内容包括安全系统工程、事故防治原理、应急管理及安全法规等相关内容。

本教材的出版，是高校安全工程专业教师践行总体国家安全观的具体生动实践，将对安全工程专业的教育发展及培养具有安全发展理念、国际视野、跨文化交流与合作能力安全工程专业人才起到积极的推动作用，同时通过编者、教师和学生的教学实践及广大读者的共同努力，本教材将得到更好的完善，为培养适应我国社会经济发展所需的安全工程专业人才贡献更大的力量。

中国安全生产科学研究院院长

长江学者特聘教授

2022年5月

前 言

安全是人类生存最重要和最基本的要求，是人类在生产过程中，将系统的运行状态对人的生命、财产、环境可能产生的损害控制在人不感觉难受的水平以下的状态。安全生产是预防生产过程中发生人身、设备事故，形成良好劳动环境和工作秩序而采取的一系列措施和活动，是安全与生产的统一。安全促进生产，生产必须安全，安全是生产的前提条件，没有安全就无法生产。

党中央、国务院历来高度重视安全生产工作，特别是党的十八大以来，以习近平同志为核心的党中央作出一系列重大决策部署，推动安全生产工作取得了历史性成就。当前，我国进入高质量发展阶段，与此同时各类安全事故不断发生，给人民的生命和财产造成巨大损失，暴露出安全发展理念不牢固的问题。如何有效消除和减少安全事故、促进安全发展已成为广大安全从业人员特别是安全科技工作者和安全人才培养教育工作者面临的问题。

安全和发展相互依存又相互制约，安全发展贯穿于我国社会发展各领域全过程，解决发展中存在的安全问题，实现安全发展，要紧紧依靠安全科技进步和安全人才培育。作为安全人才培养主力军的普通高等学校安全工程本科专业，培养了大批适应国家社会经济发展需要的安全工程专业人才。

贵州大学创建于1902年，贵州大学安全工程专业源于1958年原贵州工学院设置的矿山通风教研室，2003年经教育部批准开设安全工程本科专业，2004年招生，2007年入选贵州大学第一类特色专业，2020年入选贵州省省级一流学科建设点专业，2022年入选国家级一流专业，并根据教学大纲开设《安全工程专业英语》课程。编者根据多年安全工程专业的教学经验，结合安全工程目前及未来的发展动向进行遴选和整合，编写了这本《安全工程专业英语》教材。

本教材按照安全工程专业的培养目标，编写了18个单元，教材中所有单元均节选自国内已出版的安全工程专业英语教材、相关教辅和网上公开的相关资料。其中涉及矿山安全的有矿井通风、矿井火灾、瓦斯及控制、矿山行业中的事故分析及矿井事故防治及控制等；涉及安全基础理论的单元包括安全系统过程、故障树分析技术简介、事故防治原理、事故调查及危险辨识等；非矿山安全的相关内容兼顾有建筑安全、应急管理及化学危险等。

本教材出版获贵州大学安全工程省级一流专业建设项目资助，感谢中国安全生产科学研究院院长、长江学者特聘教授周福宝老师对本教材编写的指导和支持，感谢贵州大学矿业学院领导对安全工程专业英语课程的关注和重视。教材中部分单元的单词及词汇梳理由贵州大学安全科学与工程专业2020级硕士生李照平同学、张锐同学及2018级本科生易双霞同学协助完成；教材中的所有插图由贵州大学2019级本科生黎明银同学和赵苏州同学绘制；教材的校对和复核工作由2020级硕士生李照平同学、张锐同学及2018级本科生易双霞同学共同协助完成。对上述同学在教材编写过程中的辛苦付出表示衷心的感谢。

本教材可供大专院校安全工程专业师生作为参考资料使用，也可供具有一定安全专业知识的专业技术人员使用。

最后，特别感谢本教材参考文献的作者们，正是他们辛苦的工作和优秀的成果激发了更多的学生和专业人士对安全工程专业的兴趣与热爱，并推进安全工程专业蓬勃发展。

尽管竭尽全力，但本教材难免会存在不妥之处，恳请广大读者批评指正。

编者

2022年4月

Contents

Unit One
Mine Ventilation   1

Unit Two
Methane and Its Control   10

Unit Three
Mine Fires   24

Unit Four
Fire Hazards in Industry   29

Unit Five
Accident Prevention Principles   35

Unit Six
Hazard Identification   39

Unit Seven
Accident Investigations   45

Unit Eight
Accident Analysis in Mine Industry   51

Unit Nine
Mine Accident Prevention and Control   57

Unit Ten
The System Safety Process   64

Unit Eleven
Introduction to the Safety System Engineering   70

Unit Twelve
Introduction to Fault Tree Analysis   81

Unit Thirteen
Chemical Hazard Communication 88

Unit Fourteen
Ergonomics Engineering 95

Unit Fifteen
Health and Safety Regulation 103

Unit Sixteen
Emergency Management 109

Unit Seventeen
Construction Safety 118

Unit Eighteen
Safety Management 124

参考文献 130

Unit One

Mine Ventilation

The two purposes of mine ventilation are: (1) to answer the requirements of the law in regard to supplying a stated quantity of fresh air per minute to each man in the mine, and to dilute render harmless, and sweep away dangerous gases. In coal mines the quantity of fresh air prescribed is generally from 100 to 150cu.ft/min/man in the mine. Some mining regulations specify a maximum limit to the quantity of methane permitted in the return air of coal mines, and some limit the amount of carbon dioxide permissible in the mine air; (2) to make working conditions more comfortable for miners. If conditions of humidity and air temperature are favourable, a decisive cooling effect on the men is secured by giving the proper velocity to the current, and the efficiency of the miners is increased. Dust and fumes from explosives are also removed.

Natural and Artificial Ventilation

Pressure differences required to cause air flow, may be produced by natural or mechanical forces. Flow caused by unequal densities or weights of air columns in or near the openings (due mainly to temperature differences) is "natural-draft" flow, and resulting pressure-differences are the "natural draft pressure". The relatively feeble currents forming complete flow-circuits in undivided single openings, also due to equal densities, are separately termed "convection currents". Many metal mines and some small coal mines are ventilated by natural draft alone, which also acts in conjunction with fan pressure in mechanically-ventilated mines; Where its importance largely depends on the depth of workings and mine resistance.

The effect of natural conditions in creating a circulation of air in a mine is illustrated in Fig 1. It will be assumed that the temperature of the air current at any point in the mine is T_1, and the outside temperature is T_2. The column of air whose weight tends to produce circulation is H_1 for the main shaft, and H_2 for the air shaft. H_2 is composed of two sections, namely H_n+H_m. The direction in which the air will circulate and the pressure producing circulation may be derived by calculating. The difference between the weights of the two columns is the pressure in pounds per square foot that produces circulation of the air, and the direction of flow will be toward the column of lesser weight as indicated by the arrows in the figure.

In mines where the natural ventilation pressure is inadequate to supply the necessary air, fans are used. However, the effect of natural ventilation on the performance of the fan is important. Owing to the change in temperature from summer to winter conditions, natural ventilation may reverse its direction; in one case it assists the fan, and in the other case it opposes it.

Ventilation of coal mines is nowadays almost universally effected by the use of the fans, of which there are many types. Such fans may either exhaust the air from the upcast shaft or blow or force the air down the downcast shaft. With few exceptions, exhausting fans at the top of the up-cast shaft are used in modern mines.

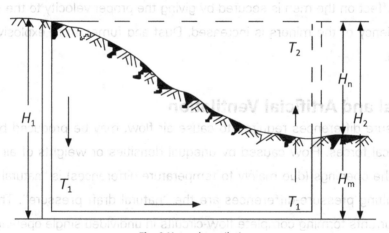

Fig. 1 Natural ventilation

Although many types of fans are used for mine ventilation, they fall into two classes, viz, the centrifugal or wheel-type fan and the axial-flow or propeller type fan.

During recent years the centrifugal fan has found a rival in the axial-flow or propeller-type fan, which is now being used in increasing numbers to such an extent that it is largely replacing the centrifugal fan for mine ventilation.

The action of the axial-flow fan differs from that of the centrifugal fan in that the

→ Mine Ventilation

air passes axially along the fan instead of being discharged from the circumference of the fan by centrifugal force. The fan consists essentially of one or more rotors (somewhat similar to aeroplane propellers; in the first axial-flow mine fans that rotors were actually aeroplane propellers). There rotors carry blades and rotate at a high speed within a circular casing which the air enters at one end and is discharged at the other end. The number of rotors or stages depends upon the pressure to be produced, and mine fans may have anything from one to four stages, with the equivalent number of rotors mounted on the same shaft.

Although apparently simple in construction and operation, this type of fan calls for a high degree of skill in the design and arrangement of the blades. With the axial-flow fan it is possible to vary the performance by increasing its speed, by increasing the number of stages or rotors, and by altering the pitch or inclination of the blades, and these alterations can be made over fairly wide limits without seriously reducing the efficiency at which the fan works.

Underground Fans

Fans are used underground mainly for two purposes, viz, as boosters for assisting the main fan, and as auxiliary fans for the ventilation of headings and blind ends.

The use of booster fans underground is confined to cases where the workings have extended to such great distances from the pit-bottom that the surface fan is incapable of circulating the quantity of air necessary for the ventilation of these remote workings and where it would be necessary either to install a larger and more powerful surface fan or to enlarge the roadways or provide additional airways to allow adequate ventilation.

Such fans are usually installed in the return airways, but when electrically driven the driving motor must be supplied with fresh or intake air.

The Distribution of Mine Ventilation

The present-day practice is to split the air near the bottom of the downcast shaft into several intake airways, each of which serves a certain area of the workings or district of the mine. Similarly, separate returns are provided for the several working areas or districts near the upcast pit-bottom.

Splitting the air in this way is essential if the large volumes of air required in modern mines are to be provided, and in addition it offers many advantages, the chief of which are:

(1) Each district is supplied with fresh air.

(2) A much large quantity of air circulates in the mine, due to lower resistance by multiple circuits or roadways.

(3) There is less risk of the accumulation of gas.

(4) In the event of trouble in a district or an explosion, the trouble or damage is more likely to be confined to the particular district in which it occurs and less likely to affect the whole mine.

(5) The velocity of the air currents in the intakes, returns and workings is lower, and the ventilating pressure required for a given total quantity of air is reduced, with consequent economy in power consumption.

Left to its own devices, the air would simply pass down the downcast and take the nearest way to the upcast shaft, leaving the rest of the mine unventilated. To prevent this and ensure the proper distribution of the air throughout the mine, various devices are employed.

Stoppings. As the mine workings advance, various connections between the intake and return airways must be sealed, as must also be abandoned roadways in order to prevent air leaking and circulating in areas where it is no longer required. It is required that any road connecting an intake and a return airway which has ceased to be required for the working of the mine shall be effectively sealed forthwith. For this purpose stoppings are constructed to confine the air along the desired course. These stoppings are built from floor to roof and from side to side of the roadways, and are constructed in many ways.

In important positions they may be built of masonry or concrete, while at other times they may consist merely of debris packed in the roadway to a sufficient thickness to prevent the passage of air.

Doors. It is frequently necessary to prevent the passage of air along roadways which must, however, be available for persons or materials to pass. In these cases ventilation doors are employed. No less than two doors are usually inserted, so that one can remain shut at all times to prevent short-circuiting of the air which would

→ Mine Ventilation

happen if a single door was used. In important situations near the pit-bottom and between main intakes and returns, it is customary to erect three or more doors, and in up-to-date mines these are sometimes constructed of steel plates with rubber beading around the edges to reduce leakage to a minimum. In other situations strong wooden doors with door frames built in brickwork surrounds are employed.

Sheets. Near the working faces, where the ventilating pressure is small and the ground is unsettled, sheets are sometimes employed as substitutes for doors to divert the air current. These consist of long brattice cloth or sacking, made windproof and usually fireproof, hung from roof to floor, and nailed to a piece of timber, often a roof bar. They can thus be lifted or pushed out of position for men or tubs to pass. The use of sheets is not recommended in positions where it is possible to insert doors, as they are far from leakproof and are easily deranged, when they allow the air to short-circuit and rob the working places of ventilation.

Air crossings. To ensure the supply of air to all parts of the mine, it frequently becomes necessary that an intake airway and a return airway shall cross each other. In such cases an air-tight bridge, called an air crossing, overcast or cross over has to be constructed.

Regulators. In order to obtain the desired distribution of air between the various districts, it is usually necessary to restrict the amount of air flowing into certain districts which offer a low resistance to air flow. This is effected by the use of regulators. It is obvious that without regulators large volumes of air would tend to flow in the splits of low resistance, leaving only small quantities for the remote workings which offer a high resistance.

A regulator usually consists of a small sliding door or an adjustable shutter set in an ordinary ventilating door.

Words and Expressions

mine ventilation 矿井（山）通风
law [lɔː] n. 规程；法令（规）
in regard to（of）关于
state [steɪt] v. 规（指、确）定；指出；表明
dilute [daɪˈluːt] v. 稀释；冲淡
render [ˈrendə(r)] v. 提供；使变得
sweep [swiːp] v. 扫（清）除；吹去（走）；排除
prescribe [prɪˈskraɪb] v. 规定；指示（令）
regulation [ˌreɡjuˈleɪʃn] n. 调节（整）；控制；规定（程）
specify [ˈspesɪfaɪ] v. 规（指、确、给）定；载明
methane [ˈmiːθeɪn] n. 甲烷；沼气
carbon dioxide [ˌkɑːbən daɪˈɒksaɪd] 二氧化碳
comfortable [ˈkʌmftəbl] adj. 舒适的
humidity [hjuːˈmɪdəti] n. 湿度
favourable [ˈfeɪvərəbl] adj. 有利的；良好的；合适的
decided cooling effect 明显的冷却作用（效果、影响）
secure [sɪˈkjʊə(r)] v. 获得；得到；保证
dust and fumes 粉尘和烟雾
artificial ventilation 人工通风；机械通风
press differences 压差
density [ˈdensəti] n. 密度；密集
air column [eə(r) ˈkɒləm] 空气柱
natural-draft [ˈnætʃrəl drɑːft] 自然通风
natrual-draft flow 自然通风（风）流
natrual-draft pressure 自然通风（风）压
feeble [ˈfiːbl] adj. 微弱的；轻微的
convention currents 对流
alone [əˈləʊ] adj/adv. 单独；仅仅；只有
in conjunction with 连同；和……在一起
mechanically-ventilated mines 机械通风的矿井
mine resistance 矿井通风阻力；矿井阻力
create [kriˈeɪt] v. 引起；产生；造（形）成
circulation of air （井下）受控风流
illustrate [ˈɪləstreɪt] v. 图解；举例说明
tend to 会；势必；趋向于……
derive [dɪˈraɪv] v. 获得；导出；推导出
calculate [ˈkælkjuleɪt] v. 计算；预测
indicate [ˈɪndɪkeɪt] v. 表（指、显）示；指出
inadequate [ɪnˈædɪkwət] adj. 不充分的；不足的
performance [pəˈfɔːməns] n. （运转）性能；特性（曲线）
owing to 由于；因为
reverse [rɪˈvɜːs] v. 反向；倒（反）转
assist [əˈsɪst] v. 辅（协、帮）助
oppose [əˈpəʊz] v. 反对；与……对抗
universally [ˌjuːnɪˈvɜːsəli] adv. 普遍地；一般地
effect [ɪˈfekt] v. 实现；完成
exhausting fan 抽出式通风机
with few exceptions 极少例外
fall into 归入；属于；分成

Unit One

viz=videlicet [vɪˈdelɪset]（拉丁）即；就是（说）

centrifugal fan/wheel-type fan 离心式通风机

axial-flow fan/propeller-type fan 轴流式（螺旋桨式）通风机

In simple terms 简单说；简言之

blade [bleɪd] *n.* 桨叶；叶片

vane [veɪn] *n.* 轮叶；叶片

revolve [rɪˈvɒlv] *v.* （使）旋转；转动

instead [ɪnˈsted] *adv.* 而是；代之以

throw [θrəʊ] *v.* 抛；投射

be thrown outward 向外抛出

casing [ˈkeɪsɪŋ] *n.* 外（机）壳

fly off 飞离（出、脱）

sling [slɪŋ] *n.* 吊索（绳、链）；抛掷装置

displacement [dɪsˈpleɪsmənt] *n.* 位移；移置

circumference [səˈkʌmfərəns] *n.* 周边；四周

inlet [ˈɪnlet] *n.* 进（入）口；进气口

in turn 依次；顺序地

keep up （使）保持下去；维持

by means of 借助于；通过

fan drift 通风机风硐；通风机引风道

suck [sʌk] *v.* 吸入

suck up 吸取（收）

rarefy [ˈreərɪˌfaɪ] *v.* 使变稀薄

upcast air 回风（流）

less dense 不太稠密（的）

descend [dɪˈsend] *v.* 下行（降、来）

ultimately [ˈʌltɪmətli] *adv.* 最终地；终究

chief feature 主要部（零、器）件

rarely exceed 很少超过

depend on 依靠；依……而定

it is assumed that 假定；人们认为

rival [ˈraɪv(ə)l] *n.* 竞争者；对手

extent [ɪkˈstent] *n.* 程度；范围

instead [ɪnˈsted] *adj.* 代替

instead of 代替

action [ˈækʃ(ə)n] *n.* 作用；动作；运转

differ from 不同于；与……有区别

centrifugal force 离心力

consist of 由……组成

rotor [ˈrəʊtə(r)] *n.* 转子；叶轮

somewhat [ˈsʌmwɒt] *adv.* 有点；多少；稍微

aeroplane propeller 飞机螺旋桨

blade [bleɪd] *n.* 叶片；桨（轮）叶

rotate [rəʊˈteɪt] *v.* 旋转；转动

circular casing 圆形的机壳

stage [steɪdʒ] *n.* 级；段

anything from... to 从……到……的

equivalent [ɪˈkwɪvələnt] *adj.* 相（同）等的

shaft [ʃɑːft] *n.* （传动、旋转）轴

apparently [əˈpærəntli] *adv.* 表面上（看来像）；显然

call for 要求；需要

skill [skɪl] *n.* 技巧（能、艺）；（特殊）技术

Words and Expressions

arrangement [əˈreɪndʒmənt] *n.* 排列；布（配）置；结构
performance [pəˈfɔ:məns] *n.*（运转）性能（效能）
fairly wide limit 相当广大的范围
seriously [ˈsɪərɪəsli] *adv.* 严重地
booster [ˈbu:stə(r)] *n.* 辅助装置；增压器；辅助通风机
headings [ˈhedɪŋs] *n.* 巷道端头；平行掘进
blind end 不通的巷道端头
be confined to 局限于；被限制在
workings [ˈwɜ:kɪŋz] *n.* 井下巷道
pit-bottom 井底；井底车场
incapable [ɪnˈkeɪpəbl] *adj.* 无能力的；不能的
incapable of 没有……的能力；不能
remote [rɪˈməʊt] *adj.* 遥远的；远距离的
install [ɪnˈstɔ:l] *v.* 安装；建立
adequate [ˈædɪkwət] *adj.* 足够的；充分的；满足要求的
driving motor 驱动电动机
distribution [ˌdɪstrɪˈbju:ʃn] *n.* 分配（布、派）
present-day 现代（今、在）的
split [splɪt] *v.* 分（裂、开、离、割）；*n.* 风流分支
serve [sɜ:v] *v.* 为……服务；供给（应）
similarly [ˈsɪmələli] *adv.* 同样地
separate returns 单独（个别）的回风巷道

workings areas 采区
district [ˈdɪstrɪkt] *n.* 区段；采区
rejoin [ˌri:ˈdʒɔɪn] *v.* 重新聚（汇）会
in addition 此外；另外
provide [prəˈvaɪd] *v.* 提供
offer [ˈɒfə] *v.* 提供；贡献
supply [səˈplaɪ] *v.* 供给；提供
circulate [ˈsɜ:kjəleɪt] *v.* 循环；流通
multiple circuits 并联的（风）路
risk [rɪsk] *n.* 危险
event [ɪˈvent] *n.*（偶然）事件
explosion [ɪkˈspləʊʒ(ə)n] *n.* 爆炸
damage [ˈdæmɪdʒ] *n.* 事故；故障；危害
likely [ˈlaɪkli] *adv.* 或许；可能；多半；大概
confine [kənˈfaɪn] *v.* 限制（在……范围内）
occur [əˈkɜ:(r)] *v.* 发生；出现
consequent [ˈkɒnsɪkwənt] *adj.* 跟着发生的；必然的
consumption [kənˈsʌmpʃn] *n.* 消耗（量）
leave... to its own devices 让……自行设法（做）；对……不加干涉
stoppings [ˈstɒpɪŋz] *n.* 风墙；隔壁
seal [si:l] *v.* 封闭
abandoned [əˈbændənd] *adj.* 废弃的
leak [li:k] *n./v.* 漏气；渗漏
no longer 不再；已不
cease [si:s] *v./n.* 中（停）止
forthwith [ˌfɔ:θˈwɪθ] *adv.* 立即；立刻

Unit One

masonry [ˈmeɪsənri] *n.* 砖石（建筑、工程）
merely [ˈmɪəli] *ad.* 仅仅；只；不过
door [dɔː(r)] *n.* 风门
available [əˈveɪləb(ə)l] *a.* 便利的；可利用的；适用于……的
insert [ɪnˈsɜːt] *v.* 嵌（插、镶）入
shut [ʃʌt] *v.* 关闭
short circuit 短路
customary [ɪnˈsɜːt] *adj.* 通常的；习惯的
erect [ɪˈrekt] *v.* 建（设）立；安装
up-to-date 最新的；现代的
rubber beading 橡胶垫圈
situation [ˌsɪtjuˈeɪʃ(ə)n] *n.* 位置；地点；情况
sheet [ʃiːt] *n.* 风障；风幕
unsettle [ʌnˈset(ə)l] *v.* 不稳定
substitute [ˈsʌbstɪtjuːt] *n.*（以……）代替（物）
divert [daɪˈvɜːt] *v.*（使）变换方向；转移
brattice [ˈbrætɪs] *n.* 风障（帘）
sacking [ˈsækɪŋ] *n.* 粗麻布；麻袋（布）
windproof [ˈwɪn(d)pruːf] *a.* 不透风（气）的
nail [neɪl] *v.*（用钉）钉住（牢）
roof bar 顶梁
recommend [ˌrekəˈmend] *v.* 推荐；介绍；建议

far from 远远不；完全不；极不
leakproof [ˈliːkpruːf] *a.* 密闭的；防（止）漏风的
derange [dɪˈreɪndʒ] *v.* 扰（弄、打）乱；重新安排
rob [rɒb] *v.* 掠夺；夺取
air crossing/overcast 风桥
cross each other 相互交叉
ensure [ɪnˈʃʊə(r)] *v.* 确保；保证
cross-over 跨接；交叉
air-tight 不漏气的
regulator [ˈregjuleɪtə(r)] *n.* 调节风门；风窗
restrict [rɪˈstrɪkt] *vt.* 限制
effect [ɪˈfekt] *v.* 实现；完成；达到
tend to 趋向于
sliding door 滑门；活动门
adjustable shutter set 可调节的闸门装置
ordinary [ˈɔːdnri] *a.* 普通的；平常的
split into 分为……（部分）
in the event of 如果……发生
for... purpose 为……目的
be built of 由……材料建成
not less than 不超过（多于）

Unit Two

Methane and Its Control

Methane and respirable dust are the two common problems encountered in underground coal mining. They are more severe in modern longwall mining because of high production.

Methane and Its Drainage

Once the air enters the mine shaft, its composition changes, and it becomes mine air. Most notably, the dust and hazardous gases will increase and dilute the concentration of oxygen. In addition, the air temperature, humidity, and pressure will all change. When those changes occur slightly, the mine air, which is not significantly different from the atmospheric air, is called fresh air. This usually refers to the air before passing through the working faces. After passing through the working face or gob it is called the return air.

In general, mine gas refers to all the hazardous gases in mines. The most frequently encountered hazardous in underground coal mines are methane(CH_4), carbon dioxide(CO_2), carbon monoxide(CO), sulphur dioxide(SO_2), hydrogen sulfide(H_2S), nitrogen dioxide(NO_2) and hydrogen(H_2).

Methane or marsh gas, by miners it is termed firedamp or simply "gas", is the major component of the hazardous gases in underground coal mines. It occupies approximately 80%~96% by volume. Thus normally when one speaks of mine gas, one means methane. It is colorless and odorless; its diffusivity is about 1.6 times that of air. Since it has a low specific gravity (0.554), methane is easily accumulated near the roof of the roadway and working faces. Though it is harmless to breathe in small quantities, it is suffocating if its concentration is very high.

The most dangerous problem with methane is the potential of methane explosion. It will be ignited when its concentration is between 5% and 16% (9.5% is the most dangerous) and the air temperature is from 1,200 to 1,382 °F (650~750 °C). Some coal seams and rock strata contain large amounts of methane, and under high pressure, the coal and gas will burst out suddenly and simultaneously. Obviously, certain appropriate measures must be employed to extract methane from these coal seams in advance.

The amount of methane emission in an underground coal mine can be expressed either by the absolute amount or the relative amount of emission. The absolute amount of emission is the absolute amount of emission per unit time in the whole mine. Its volumetric unit will be ft^3/day (m^3/day) or ft^3/min (m^3/min). However, the relative amount of emission is the average amount of emission per ton coal produced within a certain period of time, ft^3/ton (m^3/ton).

During a normal production period the methane concentration is diluted to below the lowest limit allowed by law mainly adjusting the volume of the ventilated air. The required volume of air in a working face can be determined by $Q_{air} = (Q_{gas}/c)*k$; where Q_{air} is the required fresh air volume in ft^3/min (m^3/min), and c is the maximum allowable limit of methane concentration in the return air, generally 1%~1.5%. The allowable limit of methane concentration varies from country to country. For instance, the limit in China is 1%; Holland 1.5% up to 2% in some area; Germany 1%~1.5%; France 1.5%~2% for some faces with monitoring instrument; and in the U.S. 1%~2%. K is the nonuniform coefficient of gas emission, generally 1.5.

Following the recent rapid development in longwall machinery, the longwall productivity has improved greatly while the coal produced is much smaller in size. These two events increase the amount of methane emission tremendously and consequently require a much larger volume of ventilated air. For example, in the United States the fresh air required at the longwall face is from 1.800 to 50.000 ft^3/min (510~1.417 m^3/min).

Most of the methane produced during coalification and metamorphism escapes to the atmosphere through fissures in the strata. A small part stays in the fissures in the surrounding strata and still another small part remains in the coal. The methane stays in the coal or the fissures in the surrounding strata either in a free or adsorbed state. The free methane moves freely in the coal or the fissures and fractures in the strata, whereas the gas molecules in the adsorbed methane tightly adhere to the surface of the interior fissure or the interior of coal particles. Under certain conditions, the free and adsorbed states are in equilibrium. As the pressure, temperature, and mining conditions change, the equilibrium will be destroyed. When the pressure is increased or the temperature is decreased, some parts of the free methane will become adsorbed. Conversely, some of the adsorbed methane will be released to become free methane. During mining operation the coal seams and the surrounding strata are subjected to continuous fracturing, which increases the passageways for the methane and destroys the equilibrium between the free and adsorbed methane that exists under natural conditions. As a result, some of the adsorbed methane will be freed. Thus under normal conditions, as mining progresses, the methane in the coal and the surrounding strata will be released continuously and uniformly. This is the basic form of methane emission. Only methane in the free gas state can flow into mine workings.

The methane content of a seam and surrounding strata is the most important factor controlling the amount of methane to be emitted. If the seam contains a large amount of methane, it will emit more methane during mining. In addition, the methane content in the coal seam and the surrounding strata also depends on the seam depth and geological conditions. Generally the methane content increases with the seam depth. If the seam is close to the surface, especially if there are outcrops, methane will escape to the atmosphere and consequently the methane content will be lower. The seam inclination is also a controlling factor. Since flowing along the bedding planes is much easier than flowing perpendicular to them, the larger the seam inclination, the more the methane escapes. If the surrounding strata are thick and tight in a structure, the methane will more likely remain in the strata. Conversely, if the fissures are well developed in the strata, the methane will escape easily.

If the seam being mined has a high methane content, the mining method employed should be those that extract with high recovery and leave as little coal in the gob as possible. In this respect, longwall mining is the most suitable.

During coal cutting the amount of methane emission increases sharply. However, different methods of coal cutting produce different amounts of methane emission. It depends mainly on the amount of coal cut loose, the size of the newly exposed coal face, and the size of the broken coal. For example, if air picks are used, the amount of methane emission increases 1.1~1.3 times; 1.4~2.0 times for blasting;1.3~1.6 times for shearer cutting; and 2.0~4.0 times for hydraulic jetting. This is why coal seams with a high methane content are not suitable for hydraulic mining. In modern longwall faces, the shearer cuts rapidly, resulting in high production. Consequently the amount of methane emission is large. It will be necessary to strengthen ventilation in order to reduce the methane concentration.

If longwall mining with the full-caving method is used, the methane originally stored in the roof strata and adjacent seams will be released and will flow into the normal ventilation networks. This is especially true during the periodic roof weighting when the main roof acts vigorously and caves in large areas. It may also reactivate the static air accumulated in the gob and flow into the face area and the tailentry. If the sealing method is used, the gob must be kept sealed tightly, because in a sealed gob, the methane accumulated may reach as high as 60%~70% in the static air. The methane-rich static air should not be allowed to leak into the normal ventilation networks if, on the other hand, the open gob is employed, the gob must be ventilated adequately to reduce the potential of accumulating high concentrations of methane in certain areas.

Methods of Preventing Methane Explosion 1

There are three requirements for methane explosion: a minimum concentration of methane and of oxygen and a suitable heat source. The minimum concentration, 5%, is the lower explosion limit, and 15% is the upper limit. If below 5%, it forms a bluish stable combustion layer around the flame without initiating explosion. If larger than 15%, there is an insufficient amount of oxygen to promote the chemical reactions leading to explosion. When the methane content in fresh air is 9.5%, once it encounters a heat source of a sufficient temperature, the whole amount of methane and oxygen will participate in the chemical reactions.

It must be noted, however, that as the oxygen content in the air decreases, the lower explosion limit will slowly increase while the upper explosion limit will drop sharply. When the oxygen content is decreased to 12%, the methane-air mixture will not be ignited. If the gob is sealed, there will be considerable accumulation of methane. But it will not be ignited even if there were spontaneous combustion in the remanent coal. This is due to the fact that in the sealed gob, there is an insufficient amount of oxygen in the air.

The ignition temperature is the lowest temperature for igniting a methane explosion and generally ranges from 1,202 to 1,292 °F (650~700°C). There are many underground heat sources that can ignite a methane explosion. These include any exposed fires, spontaneous coal combustion, electric arcing, high temperature gases from blasting, every hot metal surface and spark due to the impact and friction. However, once the methane-oxygen mixture encounters the heat source it requires a minimum reaction time before explosion. Although the reaction time is extremely short(Table), it is very important for mining operations. Therefore, when using permissible explosives, as long as the shot-firing is properly implemented, the methane will not be ignited.

Table Reaction time for methane exlosion

Methance concertration (%)	Heat source temperature °F (°C)						
	1,292 (700)	1,337 (725)	1,382 (750)	1,427 775	1,517 825	1,697 925	1,877 1,025
	Reaction times,sec.						
4	8.2	3.6	2.4	1.4			
6	10.2	4.3	2.6	1.5	0.62	0.21	0.07
8	14.0	5.2	3.0	1.6	0.67	0.25	0.08
10		6.3	3.5	1.75	0.72	0.26	0.09
12		7.9	4.1	1.90	0.77	0.27	0.09

In underground coal mines , methane explosion can occur in any place, however, most of them occur at the working faces where methane emission is the largest. Based on the factors contributing to methane explosion, the most effective methods for preventing methane explosion are to reduce the accumulation of methane and to eliminate high-temperature heat sources.

The areas where methane is likely to accumulate are the gob, working faces at the development entries, gob-side tailentry T-junction, near cutting drums off the shearer, and in the roof fall cavities.

It is very likely that methane accumulates to a high concentration in the gob. In the United States the gobs are ventilated to prevent methane accumulation and to reduce the temperature. In most other countries the gobs are tightly sealed that it completely cuts off any fresh air flowing into the gob or prevents high-concentration methane air flowing out of the gob. In any event, if the amount of methane emission is large, some methods of methane drainage directly from the gob to the surface are necessary. The withdrawn methane can be used as a fuel or as a raw material for chemical by-products.

Frequently at the working faces of the development entries, due to an insufficient air volume and speed, the methane cannot be effectively diluted and/or swept away. The methane concentration may reach a critical level. Since the specific gravity of methane is very small, it tends to accumulate near the roof line and forms a methane layer, sometimes up to 8~12 inches (200~300 mm) thick. It can be diluted or swept away by directing air flowing at 1.64~3.28 ft/ sec(0.5~1 m/sec). If necessary, a guide board or pipe, or a perforated compressive air pipe may be installed along the roof line to dilute the methane layer.

To increase the air volume and the air speed is an effective method for diluting the methane concentration in the entries. But if the methane emission is very heavy, other supplementary measures are necessary. These include: (1) Natural drainage. With this method, several entries are driven alternately. The methane will drain itself during the period of alternate stoppage; (2) Drain as advance. With this method, holes are drilled on either one or both ribs approximately 49~66 ft(15-20m) outby from the face. Each hole is connected to the drainage pipe out; (3) Holes are drilled ahead of the face and the methane is drained for a period of time before the face is advanced.

Methods of Preventing Methane Explosion 2

The tailentry corner is the major area where high-concentration methane accumulates. This is due to the facts that, first, it serves as the major exit for the high-concentration methane in the gob, and second, when the fresh air reaches the tailentry T-junction it has to make a 90° turn which results in a turbulent air flow in the tailentry corner. Consequently, the methane accumulated in this area cannot be carried away. Several methods can be employed to eliminate the problems:

(1) If the methane emission is heavier, some drainage methods are necessary at the tailentry corner (Fig.2). A steel pipe 150~300 ft (46~92 m) long is installed along the tailentry. The gob end of the pipe extends through the curtain separating the

tailentry from the tailentry corner. The methane accumulated at the corner will flow out through the pipe due to air pressure differentials. If the air pressure differential is too small, the drainage efficiency can be increased by installing a high-pressure water pipe or a compressed-air pipe alongside the steel pipe with nozzles at predetermined intervals connecting the two pipes.

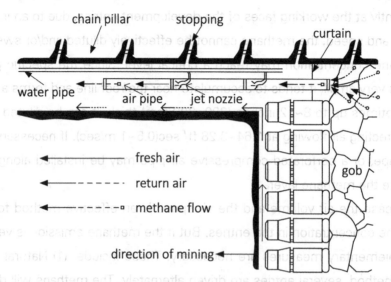

Fig. 2 Method of draining methane accumulated at the tailentry corner if the methane emission is medium high

(2) When the methane emission is larger than 176~212 ft^3/min(5~6 m^3/min), some special measures of methane drainage must be employed.

If the coal seam has a high methane content, methane emission under high production by longwall mining will be very high. In such cases, it would be rather difficult and uneconomical to dilute the methane by increasing ventilation alone. Therefore, methane drainage must be considered. Methane drainage involves drilling boreholes into the solid coal, the roof and sometimes the floor. The methane contained in the coal or rock within a radius of up to 200ft(60m), depending on the permeability, will flow into the boreholes from which the methane is vacuum-pumped, via pipelines, to the surface.

In the United states, the most common method for methane drainage in longwall mining is by surface boreholes. Before the retreat mining begins, one to three surface boreholes, depending on the panel length, are sunk along the centerline of the panel.

→ Methane and Its Control

Each borehole is sunk to a depth near the roof of the coal seam. The first borehole is usually located approximately 500 ft(155m) from the panel setup room. Methane begins to emit from the borehole when the longwall face reaches to a few meters within the borehole. The initial methane flow rate is high but erratic. It becomes stabilized after nearly 60 days. It is not uncommon that using this method the total methane flow reaches 1,000,000 ft^3/day and the methane emission from the gob is reduced by more than 50%.

Another gob degasification method for advancing longwall panel where methane emission of up to 3,000 ft^3/ min (85 m^3/min) per ton of coal is liberated, is shown in Fig.4.4 inches (10 cm) holes are drilled into the roof from the return entry at an angle of 60° for about 90 ft deep and at 75-90 ft (23-27 m) intervals. Bottom holes are also drilled at an angle of 60° to stay under and ahead of the faceline. All holes are fitted with 4 inches (10 cm) pipe and packed. The methane is vacuum-pumped to the surface and released into the atmosphere. This method can also be applied to retreat longwall panels with multiple entries, except that the holes will have to be drilled from the second entry.

(3) Water infusion involves drilling in seam horizontal holes into the solid coal ahead of mining. High pressure water from 300 to 1,500 psi is injected into the boreholes. The high-pressure water moves away in a cylindrical water front. As the water moves away from the borehole, the methane is also driven away. In order to prevent water leakage and to increase the infusion zone, hole is generally either grouted or sealed with packers at 5 ft (1.5 m) intervals. In general the infusion zone is approximate twice the length of the grouted portion of the hole. Therefore, with proper

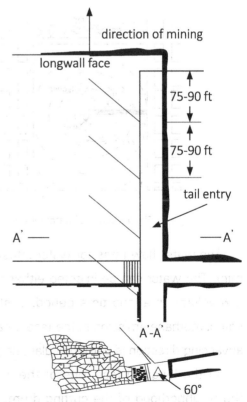

Fig. 3 Gob degasification method for advancing longwall

orientation and spacing of boreholes, the advancing water fronts can be merged to form a complete seal which in effect prevents the methane from being emitted into the coal face. In addition, water infusion tends to wet the coal before it is broken by the cutting machine. This is a very effective way of reducing the respirable dust level.

Fig.4 (A) shows a longwall retreating panel using one hole for water infusion. The panel width is 500 ft (152 m). The infusion hole is 275 ft (84 m) long. A plastic pipe 255 ft (78 m) long is inserted into the hole, with the outer 225 ft (69 m) grouted. This leaves a 50 ft (15 m) open section at the bottom of the hole for water infusion. With this arrangement the infusion zone can cover the whole face width. The infusion holes along the panel length direction should be spaced at less than 400 ft (122 m) so that the infusion zones will merge to form a complete seal.

Alternatively, two short holes, one from each side of the panel, can be drilled for water infusion in order to avoid the difficulties associated with long horizontal-hole drilling (Fig.4).

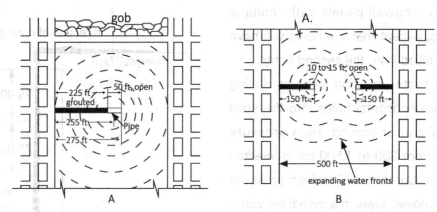

Fig. 4 Method of water infusion in a longwall panel with one(A) and two(B)

The water flow rates for water infusion range from 7 to 20 gal/min (26~76 liters/min). The water can be injected either by the high volume in a short time period or low volume in a long time period. Test results have shown that in the Pittsburgh seam, methane emission in the face area is reduced by 79% and 39% when the face advancing direction is perpendicular and parallel, respectively, to the face cleat.

At the longwall face, one of the areas where the methane accumulates is in the neighborhood of the cutting drum. This occurs because the shearer moves at a high speed, cutting a large volume of coal. Additionally the coal blocks cut loose from the face are further broken into smaller pieces as they travel rearward along the spiral vanes or scrolls of the cutting drum. As a result, an extremely large volume of methane is emitted from the coal face and the broken coal. The methane concentration is high, sometimes reaching up to 75%. Furthermore, a dead-end corner always forms in front of the leading drum where the air becomes turbulent and methane accumulates. As the web width increases, the ventilation of the dead-end corner becomes more difficult, resulting in more methane accumulation. Under this condition, if the bits cut into some rock partings, the high temperature sparks produced could ignite the methane.

The problem of eliminating methane accumulation near the cutting drum must be considered in conjunction with dust suppression methods. The simplest method for diluting the methane accumulated in this area is to increase the air volume and the air speed. If necessary, a water spray device or a small fan can be installed at the cutting drum. Furthermore, in the longwall faces with a larger volume of methane emission, the web width and haulage speed should be properly reduced so that the methane will be emitted slower and more uniformly from the face. In addition, methane emission can further be reduced by increasing the size of the broken coal.

Another important measure for the prevention of methane explosion is the strict control of high temperature heat sources. In the United States, the major heat source for the 209 cases of methane fires and explosions that occurred between 1952 and 1961 were the frictional sparks from the electrical machinery and cutting bits. Nowadays, all electrical installations are explosion-proof; that is, transformers and other electrical devices, capable of producing spark arcing, or very hot surfaces are enclosed in an antiexplosion enclosure.

Words and Expressions

respirable [ˈrespɪrəbəl] adj. 可呼吸的
encounter [ɪnˈkaʊntə(r)] v. 遇到；碰到
severe [sɪˈvɪə(r)] adj. 严重的；困难的
drainage [ˈdreɪnɪdʒ] n. 排放（泄）
composition [ˌkɒmpəˈzɪʃ(ə)n] n. 成分；组成
mine air 矿内空气；井下空气
notably [ˈnəʊtəbli] adv. 显著地；值得注意地
dust [dʌst] n. 粉尘；尘埃
hazardous gases 危险气体；有害气体
concentration [ˌkɒns(ə)nˈtreɪʃ(ə)n] n. 浓度；集中度
humidity [hjuːˈmɪdəti] n. 湿度；湿气
oxygen [ˈɒksɪdʒən] n. 氧气
slightly [ˈslaɪtli] ad. 轻微地；微小地
significantly [sɪɡˈnɪfɪkəntli] ad. 较大地；显著地
mine gas 矿井瓦斯
refer [rɪˈfɜː(r)] v. 指（的是）；涉及
marsh gas 沼气；甲烷
firedamp [ˈfaɪədæmp] n. 瓦斯；沼气
component [kəmˈpəʊnənt] n. 成分；组成部分
normally [ˈnɔːməli] ad. 正常地；通常
odorless [ˈəʊdələs] adj. 没有气味的
diffusivity [ˌdɪfjuːˈsɪvɪti] n. 扩散
specific gravity 比重
suffocating [ˈsʌfəkeɪtɪŋ] adj. 令人窒息的；呼吸困难的
ignite [ɪɡˈnaɪt] v. 点燃；着火

burst [bɜːst] v. 猝发；突（喷）出
suddenly adv. 突然地；急速地
simultaneously [ˌsɪm(ə)lˈteɪniəsli] adv. 同时发生地；同步地
appropriate [əˈprəʊpriət] adj. 适当的；恰当的
extract [ˈekstrækt] v. 抽出；排出
in advance 预先；事先
emission [ɪˈmɪʃ(ə)n] v. 放出；排出（物）
methane emission 沼气泄出
express [ɪkˈspres] v. 表示（达、明）
volumetric [ˌvɒljʊˈmetrɪk] adj. 容积的；容量的
allowable [əˈlaʊəbl] adj. 可允许的；许可的
Holland [ˈhɒlənd] n. 荷兰
Germany [ˈdʒɜːməni] n. 德国
France [frɑːns] n. 法国
monitor [ˈmɒnɪtə(r)] n./v. 监测、视、控（器）
nonuniform coefficient 不均匀系数
productivity [ˌprɒdʌkˈtɪvəti] n. 生产率（量、力）
tremendously [trəˈmendəsli] adv. 非常；巨（极）大地
consequently [ˈkɒnsɪkwəntli] adv. 因此；从而
coalification [ˌkəʊlɪfɪˈkeɪʃən] n. 煤化作用
metamorphism [ˌmetəˈmɔːfɪzəm] n. 变质作用
escape [ɪˈskeɪp] v. 逃逸；逸散；泄出
surrounding strata 围岩
adsorbed state 吸附状态

Unit Two

fissure [ˈfɪʃə(r)] n. 裂缝；裂隙
fracture [ˈfræktʃə(r)] n. 断口；断裂
adhere [ədˈhɪə(r)] v. 粘（附、固）着
interior [ɪnˈtɪəriə(r)] adj. 内部的；里面的
equilibrium [ˌiːkwɪˈlɪbriəm] n. 平衡（状态）
conversely [ˈkɒnvɜːsli] adj. 相反地；反之
release [rɪˈliːs] v. 释放出；析（逸）出
subject [ˈsʌbdʒɪkt] v. （使）经受
passageway [ˈpæsɪdʒweɪ] n. 通路；通行道
progress [ˈprəʊgres] v. 进行；前进
content [ˈkɒntent] n. 含量；内容；容量
emit [iˈmɪt] v. 喷（逸、放）出；发（放）射
outcrop [ˈaʊtkrɒp] n. 露头
bedding plane 层面
perpendicular [ˌpɜːpənˈdɪkjələ(r)] adj. （与……）垂直的；正交的
recovery [rɪˈkʌvəri] n. 回收；回采率
respect [rɪˈspekt] n. 方面；关系；（着眼）点
sharply [ˈʃɑːpli] adj. 急剧地；迅速地
air pick 风镐
hydraulic jetting [haɪˈdrɒlɪk ˈdʒetɪŋ] 水力射流冲采
strengthen [ˈstreŋθn] v. 加强；强化
full-caving method 全部垮落法
store [stɔː(r)] v. 积聚；储藏
ventilation network 矿井通风（网络、系统）
periodic roof weighting （顶板）周期来压
main roof 老顶
vigorously [ˈvɪgərəsli] adv. 强有力地

reactivate [riˈæktɪveɪt] v. （使）恢复活动
accumulate [əˈkjuːmjəleɪt] n. 集聚；积累
static [ˈstætɪk] adj. 静止（态）的
tailentry [ˈteɪlentrɪ] n. 上风巷（顺槽）
tightly [ˈtaɪtli] adv. 密封（地）；不漏地
refer to 涉及；有关；查阅
be termed 被命名为；被称为
speak of 谈到；论及
in equilibrium 平衡
result in 导致……
prevent [prɪˈvent] v. 防止；预防
requirement [rɪˈkwaɪəmənt] v. （必要的）条件；要求（物）
bluish [ˈbluːɪʃ] adj. 浅蓝色的；带蓝色的
combustion [kəmˈbʌstʃən] n. 燃烧
promote [prəˈməʊt] v. 促进；加速；助长；引起
chemical reaction 化学反应
participate [pɑːˈtɪsɪpeɪt] v. 参与（加）
initiate [ɪˈnɪʃieɪt] v. 起爆（燃）；激（触）发
remanent [ˈremənənt] adj. 剩（残）余的
spontaneous combustion 自燃
electric arcing 电弧
spark [spɑːk] n. 火花（星）
extremely [ɪkˈstriːmli] adv. 极端地；非常
permissible explosive 安全（许可）炸药
as long as 长达；只要
shot-firing 爆破（工作）；放炮
implement [ˈɪmplɪment] v. 完成；实现

Words and Expressions

contribute [kənˈtrɪbjuːt] v. 提供；促使；（对……产生）影响

effective [ɪˈfektɪv] adj. 有效的；有影响的

eliminate [ɪˈlɪmɪneɪt] v. 消（排、清）除

likely [ˈlaɪkli] adv. 或许；可能；adj. 很可能的

development entry 开拓巷道

cut off 切断

raw material 原材料

by-product 副产品（物）

sweep [swiːp] v. 扫除；吹去

critical [ˈkrɪtɪk(ə)l] adj. 临界的；极限的

specific gravity n. 比重

perforate [ˈpɜːfəreɪt] v. 穿（冲、打）孔

supplementary measure 补充（附加，辅助）的措施

natural drainage 自然排放

alternately [ɔːlˈtɜːnətli] adv. 交替地；轮流地

stoppage [ˈstɒpɪdʒ] n. 停止（顿）；停工

advance [ədˈvɑːns] v./n. 提（超）前；推进

rib [rɪb] n. 煤壁；窄煤柱

outbye [ˌaʊtˈbaɪ] adv. 在不远的地方

lead to 导致；通向

participate in 参与（加）

based on 根据

such that 如此以致；因而

ahead of 在……的前面

exit [ˈeksɪt] n. 出口；通道

turbulent [ˈtɜːbjələnt] adj. 扰动的；紊流的

corner [ˈkɔːnə(r)] n. 拐（隅）角

curtain [ˈkɜːt(ə)n] n. 风帘（障、幕）

nozzle [ˈnɒzl] n. 喷嘴（头）

differential [ˌdɪfəˈrenʃl] n. 差别；差异

rather [ˈrɑːðə(r)] adv. 相当地；颇为

radius [ˈreɪdiəs] n. 半径

permeability [ˌpɜːmiəˈbɪləti] n. 渗透（性、率）；透气性

vacuum [ˈvækjuːm] n./adj. 真空（的）

via [ˈvaɪə] prep. 经过；经由

sink [sɪŋk] v. 打进

setup room 开切眼

erratic [ɪˈrætɪk] adj. 不正常的；不规律的；不稳定的

stabilize [ˈsteɪbəlaɪz] v. 稳定；安定

degasification [dɪˌɡæsɪfɪˈkeɪ(ə)n] n. 脱（排、除）气（作用）

liberate [ˈlɪbəreɪt] vt. 放出；释出

fit [fɪt] v. 安装；装备

pack [pæk] v. 装（充）填；密封

except [ɪkˈsept] conj. 只有；除……之外

infusion [ɪnˈfjuːʒn] n. 注水；灌注；侵（渗）入

water infusion 注水

psi [psaɪ; saɪ] n. 磅/平方英寸

inject [ɪnˈdʒekt] v. 注入；灌入

cylindrical [səˈlɪndrɪkl] adj. 圆柱形的；筒形的

front [frʌnt] n. （锋、正、前）面

grout [ɡraʊt] v. 灌浆

Unit Two

packer [ˈpækə(r)] *n.* 充填（材料物）；栓塞
orientation [ˌɔːriənˈteɪʃn] *n.* 定向；校正（排列）方向
space [speɪs] *v.* 留（间隔）；配置
merge [mɜːdʒ] *v.* 消失；融合；吸收
in effect 实际上；事实上；有效
cover [ˈkʌvə(r)] *v.* 覆盖
alternatively [ɔːlˈtɜːnətɪvli] *adv.* 换句话说；相互交替地
avoid [əˈvɔɪd] *v.* 避免；取消
associate [əˈsəʊsieɪt] *v.* 伴随……（产生）
Pittsburgh [ˈpɪtsbɜːg] （美国）匹茨堡
cleat [kliːt] *n.* 层（解、节）理
faceleak 主解理面
neighborhood [ˈneɪbəhʊd] *n.* 附（邻）近；周围
additionally [əˈdɪʃənəli] *adv.* 另外；加之
rearward [ˈrɪəwəd] *adv.* 向后面（方、部）
spiral vanes 螺旋叶片
scroll [skrəʊl] *n.* 螺旋体
furthermore [ˌfɜːðəˈmɔː(r)] *adv.* 而且；此外
dead-end corner 尽头；隅（死）角

web width 截槽深度；工作面推进距离
bit [bɪt] *n.* 截齿；凿子
in conjunction with 和……一起；连同……一起
suppression [səˈpreʃ(ə)n] *n.* 抑制；消除
water spray device 喷（洒）水装置
strict [strɪkt] *adj.* 严格（密）的
installation [ˌɪnstəˈleɪʃn] *n.* （整套）装置；设备
enclosed [ɪnˈkləʊzd] *adj.* 封闭（式）的；密封的
enclosure [ɪnˈkləʊʒə(r)] *n.* （外、机）壳；罩；箱
make a turn 转弯
be carried away 清除；带上
be fitted with 装备；装上
in effect 有效地；实际上
as a result 作为……结果；结果是
under this condition 在这种情况下
capable of 能够……
be enclosed in 密封在……中

Unit Three

Mine Fires

Coal Mine Fires

Three classes of fires encountered in coal mines are:

Class a-fires in which the fuel is coal, rubber, wood or other solid burnable materials.

Class b-fires in which the fuel is a liquid, such as gasoline or oil.

Class c-fires which occur in electrical equipment during arcing.

Most coal mine fires result from electrical failures, which cause arcs and class c blazes, which normally last only a short time but are important as the ignition source of other fires. A study of 92 face-equipment fires showed that 50 occurred on cutting machines. For the 92, points of origin were: trailing cable on reel, 35; trailing cable off reel, 22; machine cable, 26; others, 9. Also notable was that 85% of the fires occurred on d-c equipment and only 7% on a-c. Further, a substantial number of fires involved ignition of coal dust or oil accumulations on the machines.

Other sources of fires include open flames, ranging from bare trolley or other wires to wood crossbars, door beaders, top coal and the like, and spontaneous combustion in gob areas where the materials (coal, broken timber, etc.) are combustible, the heat of combustion is retained because the material is a poor conductor, ventilation is sufficient to provide the necessary oxygen but not sufficient to carry away the heat of oxidation or combustion, and there is enough air to keep the fire burning after it starts.

Fire prevention—Methods of preventing mine fires include, in brief, the following:

Use of flame-resistant materials—conveyor belting and hydraulic fluid, as examples.

Good housekeeping—Keeping mine openings free of oily rags, wood, rubber and other combustible materials and storing and handling oil and other flammable materials to prevent leakage and spillage; keeping equipment free of oil, coal dust and other flammable materials.

Good maintenance from both the electrical and mechanical standpoints. No arcs in equipment or cables means no ignitions from this source. And from the mechanical standpoint, even if the hydraulic fluid is of the fire-resistant type, rupture of a hydraulic hose or leakage can leave oil on coal, increasing the liability of ignition, as a result of the evaporation of the water forming the emulsion.

Use of properly rated and installed fuses and circuit breakers to quickly cut power off face equipment and cables and thus reduce the intensity and duration of arcs.

Proper installation and guarding of trolley, feeder and other bare wire to prevent arcing to coal and timber.

Firefighting—Keeping little fires that do start in spite of the preceding and other precautions is the major objective in firefighting, approaches include:

Fire extinguishers at strategic locations and on equipment.

Fire extinguishing systems of the liquid, dry-chemical or foam type on mobile equipment, shuttle cars, miners and so on.

Installation of water lines large enough to supply sufficient water, with taps and valves every 300 to 400 ft for connecting fire hoses, plus shut off valves every 1000 ft, since experience has shown that water is the most effective means of coping with fires that are not controlled during their early stages. Federal regulations spell out the minimum requirements for a water-supply system and fire hoses for use in coal mines.

Acquisition of high expansion foam equipment for use if the fire is or spreads beyond the effective range of portable extinguishers and hose lines. If available in sufficient quantity, rock dust can serve as a fire extinguisher when conditions are right.

The last resort in fighting a fire is sealing that portion of the mine or, in severe cases, the entire mine. Flooding of sealed sections with water, where the dip makes it possible, or with nitrogen, has been done in a few instances of such steps, at least 100 days usually is allowed for extinguishment. Usually, after 100 days, an air analysis of 0.0% carbon monoxide, less than 1% oxygen and high concentrations of nitrogen and methane indicate a good possibility of success. Extreme caution must be exercised when removing seals to prevent the rekindling of the fire or the ignition of an explosive mixture in the sealed atmosphere.

Survival Measures

Carbon monoxide is the major hazard for men after explosion and during and after mine fires. Self-rescuers provide individual protection, and the act of 1969 provides that an approved unit capable of functioning for 1 hr or more be made available to each miner. If not worn on the person because it would hamper the man's activities and thus constitute a hazard, a unit must be available in a place not 25 ft faraway. In the early days of mining workers on occasion achieved protection by going into a fresh-air place and sealing it off, and some mines provided materials for such sealing at strategic spots. Now, under the act of 1969, the secretary may prescribe that rescue chambers, sealed and ventilated, be built and equipped with first-aid materials, self-contained breathing apparatus and an independent communications system to the surface. A general communication system is mandatory in all coal mines.

Underground Coal Fires—a Looming Catastrophe

Coal burning deep underground in China, India and Indonesia is threatening the environment and human life, as scientists have warned. These large-scale underground blazes cause the ground temperature to heat up and kill the surrounding vegetation, produce greenhouse gases and can even ignite forest fires, a panel of scientists told the annual meeting of the American association for the advancement of science in Denver. The resulting release of poisonous elements like arsenic and mercury can also pollute local water sources and soils, they warned.

"Coal fires are a global catastrophe," said associate professor Glenn Stracher of east Georgia College in Swainsboro, USA. But surprisingly few people know about them.

Coal can heat up on its own, and eventually catch fire and burn, if there is a continuous oxygen supply. The heat produced is not caused to disappear and under the right combinations of sunlight and oxygen, can trigger spontaneous catching fire and burning. This can occur underground, in coal stockpiles, abandoned mines or even as coal is transported. Such fires in China consume up to 200 million tones of coal per year, as delegates were told. In comparison, the U.S. economy consumes about one billion tones of coal annually, said Stracher, whose analysis of the likely impact of coal fires has been accepted for publication in the international journal of Coal Ecology. Once underway, coal fires can burn for decades, even centuries. In the process, they release large volumes of greenhouse gases, poisonous fumes and black particles into the atmosphere.

The panel members discussed the impact these fires may be having on the global and regional climate change, and agreed that the underground nature of the fires makes it difficult to protect them. One of the members of the panel, assistant professor Paul Van Dijk of the international institute for Geo-information Science and Earth Observation in the Netherlands, has been working with the China government to detect and monitor fires in the northern regions of the country.

Ultimately, the remote sensing and other techniques should allow scientists to estimate how much carbon dioxide these fires are emitting. One suggested a method of containing the fires presented by Gary Colaizzi, of the engineering firm Goodson, which has developed a heat-resistant grout (a thin mortar used to fill cracks and crevices), which is designed to be pumped into the coal fire to cut off the oxygen supply.

Words and Expressions

Unit Three

burnable [ˈbɜːnəb(ə)l] adj. 易燃的
result from 起源于；由于
d-c equipment 直流设备
substantial [səbˈstænʃ(ə)l] a. 大量的；结实的
open flame 明火
trolley [ˈtrɒli] n. 手推车；有轨电车
and the like 诸如此类
spontaneous [spɒnˈteɪniəs] adj. 自发的；自然的
combustion [kəmˈbʌstʃən] n. 燃烧；氧化
poor conductor 不良导体
oxidation [ˌɒksɪˈdeɪʃn] n. 氧化
flame-resistant 耐火的；抗火的
free of 摒除；打消……
maintenance [ˈmeɪntənəns] n. 维持；保持
evaporation [ɪˌvæpəˈreɪʃn] n. 蒸发；消失
emulsion [ɪˈmʌlʃn] n. 乳状液；感光乳剂
timber [ˈtɪmbə(r)] n. 木材；木料
precaution [prɪˈkɔːʃn] n. 预防；防备
fire extinguisher 灭火器
shut off 停止；关闭；隔绝

spell out 阐明；读出
rekindle [ˌriːˈkɪndl] v. 使再燃
carbon monoxide n. 一氧化碳
self-rescuer 自救器
individual [ˌɪndɪˈvɪdʒuəl] adj. 单独的；个人的；n. 个人；某种类型的人
on occasion 有时
apparatus [ˌæpəˈreɪtəs] n. 机构；器械
strategic [strəˈtiːdʒɪk] adj. 至关重要的；关键性的
rescue chambers 救援室
looming [ˈluːmɪŋ] adj. 迫在眉睫的
catastrophe [kəˈtæstrəfi] n. 灾难
poisonous [ˈpɔɪzənəs] adj. 有毒的
arsenic [ˈɑːsnɪk] n. 砷
mercury [ˈmɜːkjəri] n. [化] 汞；水银；[天] 水星
mortar [ˈmɔːtə(r)] n. 砂浆
grout [ɡraʊt] n. 水泥浆；石灰浆
crevices [ˈkrevɪs] n. (尤指岩石的) 裂缝；缺口

Unit Four

Fire Hazards in Industry

Fire spreads through factory premises

Two employees are killed when gas cylinder explodes

126 jobs lost as fire-hit company closes

Chemical company fails to recover after major fire

Unfortunately, these headlines are all too common. In the United Kingdom, fire brigades attend more than 40,000 fires in workplaces every year. These fires kill more than 30 people per year and injure almost 3,000 people. In addition, insurance claims for fire damage in workplaces amount to an average of £ 10 million per week.

Fire kills, injures and causes damage to property, and it can also have significant effects upon the future of companies. It is estimated that 60% of businesses that have suffered serious fire damage fail to recover and cease trading within five years of the initial incident.

The Association of British Insurers released a press statement in the late 1990s, which stated that:

Britain's businesses and their insurers are being hit by a major rise in the cost of fire damage, with losses costing over £1.6 million every day. In 1998, fire damage claims cost insurers £ 600 million, 22% up on the previous year to their highest level since 1992. The combined cost of fire damage and business interruption losses, at £ 808 million, rose by 13% on the previous year.

Although this was a stark message, an even more disturbing message was perhaps contained within a further statement:

The current level of commercial fire losses is very disturbing and needs to be tackled. These losses are unsustainable and can only put pressure on premium rates unless there is a significant improvement during the year. Businesses must recognize the urgent need to reduce the risk from fire, and make this a priority for this year.

Although accidental fires are on the increase, one of the most concerning categories of fires are those related to arson. It is estimated that fires that are started deliberately account for 40% of workplace fires. In some sectors, this figure is significantly higher. For example, in the category of recreational- and cultural-type premises, 66% of fires are started deliberately.

The risk from fire is, therefore, an important area for employers to control and manage properly. Generally, there are three main reasons for managing fire risks.

Firstly, there is a moral duty on employers to provide a safe workplace that is free from risks to health. Employees perform better at work and enjoy their working environment when they can see that their safety has been considered and that steps have been implemented to ensure that risk is controlled. In this age of modern health and safety legislation, it is not considered morally acceptable to put people at risk unless under special circumstances, e. g. emergency incidents.

Secondly, from an economic point of view, fire safety management is also important. As previously stated, fires cost significant sums of money. A common belief is that most of the damage that results from fire can be recovered from insurance companies. Although not relating specifically to fire incidents, the Health and Safety Executive carried out a study in the 1990s, whereby the true cost of accidents was investigated. As an average, it was calculated that, should an incident occur, only 8% of the loss would be recovered through insurance.

Direct costs such as property damage and personal injury are definitely recovered from insurers. However, it is the hidden costs, or the indirect costs, that are important. These costs include administration costs, loss of production, loss of orders, bad publicity as a result of accidents and possible litigation. These indirect losses may have a detrimental effect on business recovery after a serious fire. In fact, more than half of the companies experiencing a serious fire will fail to recover enough to continue trading. This figure is actually higher for small to medium-sized enterprises.

Although economic costs are important, a final reason for managing fire safety is because it is a legislative requirement. There are two main items of legislation that relate directly to fire precautions, that is, the Fire Precautions Act 1971 and the Fire Precautions (workplace) Regulations 1997 (as amended). Both of them are enforced by the local fire authorities, with the exception of Crown-occupied premises where enforcement is carried out by the Fire Services Inspectorate of the Home Departments.

The Fire Precautions Act deals with general fire precautions such as:

(1) Means of detecting a fire and giving warning in the event of a fire.

(2) The provision of a suitable means of escaping from premises that come under the Act.

(3) The provision of suitable fire-fighting equipment, including fire extinguishers, hose reels and sprinkler systems.

(4) Training of staff in relation to fire safety arrangements, e. g. evacuation procedures, fire alarm systems, fire-fighting and generally awareness of fire hazards.

In addition to the Act, the Fire Precautions (workplace) Regulations require employers to carry out fire risk assessments. Assessing the risk from fire involves the following stages:

Stage one—identification of fire hazards, including sources of ignition, sources of fuel and work activities that may present a fire hazard.

Stage two—identification of the number, location and type of persons at risk from fire, including employees, contractors, visitors and members of the public.

Stage three—evaluation of the risk from fire hazards. This stage involves evaluating the existing arrangements for controlling fire risks, such as the control of ignition sources and sources of fuel, fire detection, means of escape, the maintenance of fire controls and training of employees. Where these existing arrangements are sufficient, no further action is required. However, where the existing controls are found to be inadequate, further control measures need to be identified, implemented and maintained.

Stage four—recording the findings of the fire risk assessment. This may be done using a paper based system or by electronic means. Also, employees and others who may be affected by the fire risks need to be consulted, made aware of the risks and trained for the measures required to control the risks.

Stage five—reviewing the assessment when significant changes occur to the process or after a fixed period of time.

Words and Expressions

hazard [ˈhæzəd] *n.* 危险；危害物；*v.* 冒险做（无把握之事）；冒险做出
cylinder [ˈsɪlɪndə(r)] *n.* 圆柱体；圆筒状物；（发动机）汽缸
fail to do sth. 未能做……
brigade [brɪˈɡeɪd] *n.* 旅；伙；帮；派
United Kingdom *n.* 英国；联合王国
In addition 除此之外
insurance [ɪnˈʃʊərəns; ɪnˈʃɔːrəns] *n.* 保险（业）；保费；预防措施
claim [kleɪm] *v.* 声称；索取；*n.* 声称；主张
have significant effects upon 有显著的影响
It is estimated that 估计
suffer [ˈsʌfə(r)] *v.* 受苦；忍受
cease [siːs] *v.* 停止；终止
initial [ɪˈnɪʃl] *adj.* 开始的；*n.* 首字母；*v.* 用姓名的首字母签名于
The Association of British Insurers 英国保险公司协会
release [rɪˈliːs] *v.* 释放；松开；*n.* 释放；发行
insurer [ɪnˈʃʊərə(r); ɪnˈʃɔːrə(r)] *n.* 承保人；保险公司
interruption [ˌɪntəˈrʌpʃn] *n.* 中断；打断
stark [stɑːk] *adj.* 完全的；荒凉的；*adv.* 完全；明显地
further [ˈfɜːðə(r)] *ad.* 较 / 更远地；进一步地；*a.* 更多的；附加的；*v.* 促进；增进
commercial [kəˈmɜːʃ(ə)l] *adj.* 商业（化）的；营利的；*n.* 商业广告

unsustainable [ˌʌnsəˈsteɪnəbl] *a.* 不能持续的；无法维持的
put pressure on 强迫；促使
premium [ˈpriːmiəm] *n.* 额外费用；保险费；附加费；*adj.* 高昂的；优质的
priority [praɪˈɒrəti] *n.* 优先；优先权；重点
on the increase 正在增加之中；不断增加
category [ˈkætəɡəri] *n.* 种类；类别；派别
relate to 与……有关
sector [ˈsektə(r)] *n.* 部门；领域
figure [ˈfɪɡə(r)] *n.* 数字；位数；人影；*v.* 认为；计算；出现
moral duty 道德义务
free from （把……）从……释放出来，使摆脱
perform [pəˈfɔːm] *v.* 执行；起……作用
implement [ˈɪmplɪmənt] *vt.* 实施；执行；落实（政策）；*n.* 工具；器械；履行（契约）
ensure [ɪnˈʃʊə(r)] *v.* 确保；担保
legislation [ˌledʒɪsˈleɪʃn] *n.* 立法；制定法律
from an economic point of view 从经济的角度来看
as previously stated 如前所述
sum [sʌm] *n.* 总数；算术；*v.* 合计；总结
results from 结果来自
executive [ɪɡˈzekjətɪv] *n.* 经理；行政部门；*adj.* 行政的；管理的

Unit Four

carry out 执行；进行
investigate [ɪnˈvestɪgeɪt] v. 调查；研究
it was calculated that 据计算
hidden [ˈhɪd(ə)n] adj. 隐藏的；神秘的；v. 隐藏（hide 的过去分词）
indirect [ˌɪndɪˈkeɪt] adj. 间接的；不坦率的
litigation [ˌlɪtɪˈgeɪʃn] n. [律] 打官司；诉讼
have a detrimental effect on 对……有害的影响
in fact 实际上，其实
medium-sized enterprises 中型企业
legislative [ˈledʒɪslətɪv] adj. （关于）立法的；立法决定的
relate directly to 与……有直接联系；直接相关；涉及
precaution [prɪˈkɔːʃ(ə)n] n. 预防；防备
Fire Precautions Act 火灾预防条例
Fire Precautions Regulations 消防措施规定
amend [əˈmend] v. 修正；修订
enforce [ɪnˈfɔːs] v. 强制执行；强迫
authority [ɔːˈθɒrəti] n. 当局；权力
fire Services Inspectorate 消防监督所
Home Departments 国家部门
deal with 处理；应付
means [miːnz] n. 方法；收入
detecting [dɪˈtektɪŋ] n. 探测；检定
provision [prəˈvɪʒn] n. 规定；条项
escape from 逃避；逃出
fire-fighting adj. 消防的

fire extinguishers 灭火器
hose reels and sprinkler systems 软管卷轴和自动喷水灭火系统
staff [stɑːf] n. 全体职工；行政人群；v. 任职于……
evacuation procedures 疏散程序
alarm [əˈlɑːm] n. 警报（器）；闹钟；v. 使惊恐；使担心
In addition to 除……以外（还）
assessment [əˈsesmənt] n. 评估；评价
ignition [ɪgˈnɪʃ(ə)n] n. （汽油引擎的）发火装置；着火；燃烧
present [ˈprez(ə)nt] adj. 目前的；在场的；n. 现在；礼物；v. 呈递；导致
contractor [kənˈtræktə(r)] n. （建筑、监造中的）承包人
maintenance [ˈmeɪntənəns] n. 维持；保持；保养
sufficient [səˈfɪʃ(ə)nt] adj. 足够的；充足的
be found to 被发现
inadequate [ɪnˈædɪkwət] adj. 不充足的；不适当的
be affected by 受……影响
tank [tæŋk] n. 箱；坦克
vessel [ˈves(ə)l] n. 容器；船；血管
Association of British Insurers or ABI 英国保险公司协会或 ABI
investment [ɪnˈvestmənt] n. 投资额；（时间、精力）投入

 **Words and Expressions**

 **Unit Four**

represent [ˌreprɪˈzent] v. 表现；代表
collective [kəˈlektɪv] adj. 共有的；总体的
interest [ˈɪntrəst] n. 兴趣；利益；利息
participate in 参加；参与
act as 担任；担当
advmocate [ˈædvəkeɪt] n. 丙二酸酯
whilst [waɪlst] conj. 在……期间；与……同时；然而；尽管；n. 一会儿
rescue [ˈreskjuː] v. 营救；救助；n. 营救（行动）
prompt [prɒmpt] adj. 敏捷的；迅速的；

v. 促使；鼓励；提示；n. 激励；提示符
Factories Act 工厂法
Hotels and Boarding Houses 酒店和寄宿房屋
Statutory Instrument (SI) 法定文书
accommodation [əˌkɒməˈdeɪʃ(ə)n] n. 住宿；调解
certificate [səˈtɪfɪkət] n. 证明；证书；文凭；v. 发结业证书

Unit Five
Accident Prevention Principles

Coal mining historically has been a hazardous occupation but, in recent years, tremendous progress has been made in reducing accidental coal mine deaths and injuries.

Accident prevention fundamentals are to no different for coal mining than for any other type of work. The fundamentals apply to all industries. Construction workers may be struck by falling tools and materials, and miners may be trapped by mine fires. Railroad brakemen may lose hands and fingers positioning couplers, miners may also lose hands and fingers positioning couplers, and many other examples could be cited, but the basic point is that accidents that happen in coal mines are basically similar to accidents that happen in other industries. Thus, the basic principles for preventing accidents apply in coal mining as they do in any other type of work.

A personal accident is defined as an unexpected happening that interrupts the normal work activity of an employee and usually, although not necessarily, results in an injury. An equipment accident is an unexpected happening that results in damage to equipment and under certain circumstances results in injuries to employees.

The supervisor should understand why men act in an unsafe manner and how unsafe conditions are produced, so he will be able to recognize and correct both situations before they lead to an accident. He can learn this in many ways, but one major method is from thorough accidents.

Accidents are complex. Usually there are several causes, both personal and environmental, that operate in sequence or in combination to result in an accident.

To prevent an accident from happening, the causes of accidents must be determined and eliminated. Accidents are caused! They don't just happen.

The basic principles of accident prevention are establishing the potential and actual causes of accidents, and eliminating the causes.

Unit Five

The first principle emphasizes that causes exist before accidents happen. Properly trained supervisors can recognize potential accident causes and eliminate them before accidents happen. In the event an accident has occurred, they are able to determine the actual causes and eliminate them before they cause additional accidents.

Having determined the principles for accident prevention, what is needed to put them to work?

First, the potential and actual causes of accidents must be established.

Potential causes are the unsafe conditions and unsafe practices that have not yet caused an accident. To detect and determine potential unsafe conditions, a mine safety program should include planned safety inspections, which are explained under accident prevention tools.

There could be a tendency for coal mine operators to consider the requirements of the federal Mine safety and the health act of 1977 as the basis for an adequate inspection program. These regulations, however, are directed principally toward specific types of unsafe conditions and the physical environment. They should, therefore, not be the only items of an inspection program for unsafe conditions, but should supplement the operator's own inspection program.

To detect and correct unsafe practices that have not yet caused an accident, supervision could make regular, planned job procedure observations of a specific employee performing a specific job. This planned safety observation is also explained under accident prevention tools.

Actual conditions or actions, which have contributed to an accident, can be determined with the aid of another safety tool, the accident investigation report.

The second principle of accident prevention is to eliminate the causes of accidents.

Where the supervisor has the authority he should eliminate any unsafe conditions and any causes of the unsafe conditions. If he lacks this authority then he should report the unsafe conditions and causes to his superior together with recommendations for correction.

Unsafe practices are just as important as unsafe conditions. Unsafe practices stem from many sources, all of which must be eliminated if accidents are to be prevented.

One major cause of unsafe practices is a lack of knowledge or skill. Fortunately, it can be eliminated by job safety instructions before a man starts on a new job, by follow-up observations, and by repeated instructions of safe job procedures as explained under accident prevention tools.

Another source of unsafe practices is improper motivation and attitudes. This, too, can be eliminated by repeated emphasis on the gains to be achieved by working safely compared with the pain, suffering, and loss that are incurred in an accident. Changing attitudes require considerable tact, skills, and patience on the part of the supervisor. His knowledge of the miners, the work situation, and practices in the mine will determine how he will handle the situation.

Any safety program should be audited to determine if it is being implemented as designed. The audit of an accident prevention program should determine if each member of management is carrying out his responsibilities and if each accident prevention tool is properly used.

The audit should be a formal procedure and should be performed, if practicable, by mine supervisors who are not associated with the mine being audited.

The mine may have a well-designed safety program, but supervisors may be only going through the motions, filling in the forms, and presenting the appearance of an active safety program. If so, the accident experience might be poor. The audit determines the areas in which the program needs to be improved.

 **Words and Expressions**

 **Unit Five**

principle [ˈprɪnsəpl] *n.* 道德原则；法则；观念；理由；定律

hazardous [ˈhæzədəs] *adj.* 冒险的；有危险的；碰运气的

tremendous [trəˈmendəs] *adj.* 极大的；巨大的；可怕的；惊人的；极好的

different than 不同于

construction [kənˈstrʌkʃn] *n.* 建造（方式）；施工；建筑（物）

brakemen [ˈbreɪkmən] *n.* （火车上的）维修工（brakeman 的名词复数）；司闸员；制动手

trap [træp] *adj.* 捕集的；捕获的；陷入困境的；受到限制的；*v.* 诱骗（trap 的过去式和过去分词）；使受限制；困住；使陷入困境

similar to 跟……类似的；与……同样的；如同

result in 导致；造成

supervisor [ˈsuːpəvaɪzə(r)] *n.* 管理者；监督者；指导者

thorough [ˈθʌrə] *adj.* 详细的；仔细的；完全的

potential [pəˈtenʃl] *adj.* 潜在的；*n.* 潜力；可能性

explain [ɪkˈspleɪn] *v.* 解释；辩解；说明……的理由

supplement [ˈsʌplɪmənt] *v.* 增补；补充；*n.* 增补；补充；补充物

recommendation [ˌrekəmenˈdeɪʃn] *n.* 推荐；推荐信；建议；可取之处

as important as 同样重要

stem from 起源于

fortunately [ˈfɔːtʃənətli] *adv.* 幸好；幸运的是

motivation [ˌməʊtɪˈveɪʃn] *n.* 动力；积极性

attitude [ˈætɪtjuːd] *n.* 态度；看法

compare with 把……与……相比；可与……相比；比较

patience [ˈpeɪʃ(ə)ns] *n.* 容忍；毅力

carry out 执行；完成

Unit Six

Hazard Identification

Introductionn

Hazard identification is a process controlled by management. You must assess the outcome of the hazard identification process and determine if immediate action is necessary or if, in fact, there is an actual hazard involved. When you do not view a reported hazard as an actual hazard, it is critical to the ongoing process to inform the worker that you do not view it as a true hazard and explain why. This will insure the continued cooperation of workers in hazard identification.

It is important to remember that a worker may perceive something as a hazard, when in fact it may not be a true hazard; the risk may not match the ranking that the worker placed on it. Also , even if a hazard exists, you need to prioritize it according to the ones that can be handled quickly, which may take time ,or which will cost money above your budget. If the correction will cause a large capital expense and the risk is real but does not exhibit an extreme danger to life and health, you might need to wait until next year's budget cycle. An example of this would be when workers complain of the smell and dust created by a chemical process. If the dust is not above accepted exposure limits and the smell is not overwhelming, then the company may elect to install a new ventilation system, but not until the next year because of budgetary constraints. The use of PPE until hazards can be removed may be required.

The expected benefits of hazard identification are a decrease in the incidents of injuries, a decrease in lost workdays and absenteeism, a decrease in workers' compensation costs, increased productivity, and better cooperation and communication. The baseline for determining the benefit of hazard identification can be formulated from the existing company data on occupational injuries/illnesses, workers' compensation, attendance, profit, and production.

Hazard identification includes those items that can assist you with identifying workplace hazards and determining what corrective action is necessary to control them. These items include jobsite safety inspections, accident investigations, safety and health committees, and project safety meetings. Identification and control of hazards should include periodic site safety inspection programs that involve supervisors and, if you have them, joint labor management committees. Safety inspections should ensure that preventive controls are in place (PPE, guards, maintenance, engineering controls), that action is taken to quickly address hazards, that technical resources such as OSHA, state agencies, professional organizations, and consultants are used, and that safety and health rules are enforced.

Many workplaces have high accident incidence and severity rates because they are hazardous. Hazards are dangerous situations or conditions that can lead to accidents. The more hazards present, the greater the chance that there will be accidents. Unless safety procedures are followed, there will be a direct relationship between the number of hazards in the workplace and the number of accidents that will occur there.

As in most industries, people work together with machines in an environment that causes employees to face hazards, which can lead to an injury, disability, or even death. To prevent industrial accidents, the people, machines, and other factors which can cause accidents, including the energies associated with them, must be controlled. This can be done through education and training, good safety engineering, and enforcement.

The core of an effective safety and health program is hazard identification and control. Periodic inspections and procedures for correction and control provide methods of identifying the existing or potential hazards in the workplace and eliminating or controlling them. The hazard control system provides a basis for developing safe work procedures and injury and illness prevention training. Hazards occurring or recurring reflect a breakdown in the hazard control system.

The written safety and health program establishes procedures and responsibilities for the identification and correction of workplace hazards. The following activities can be used to identify and control workplace hazards: a hazard reporting system, job site inspections, accident investigation, and expert audits.

→ Hazard Identification

After all basic steps of the operation of a piece of equipment or job procedure have been listed, we need to examine each job step to identify the hazards associated with each job step. The purpose is to identify and list the possible hazards in each step of the job. Some hazards are more likely to occur than others, and some are more likely to produce serious injuries than others. Consider all reasonable possibilities when identifying hazards.

1. Accident Types

The Struck-against Type of Accidents

Look at the first four basic accident types—struck-against, struck-by, contact-with and contacted-by—in more detail, with the job step walk-round inspection in mind. Can the worker strike against anything while doing the job step? Think of the worker moving and contacting something forcefully and unexpectedly—an object capable of causing an injury. Can he or she forcefully contact anything that will cause injuries? This forceful contact may be with machinery, timber or bolts, protruding objects to sharp, jagged edges. Identify not only what the worker can strike against, but how the contact can come about. This does not mean that every object around the worker must be listed.

The Struck-by Type of Accidents

Can the worker be struck by anything while doing the job step? The phrase "struck by" means that something moves and strikes the worker abruptly with force. Study the work environment for what is moving in the vicinity of the worker, what is about to move, or what will move as a result of what the worker does. Is an unexpected movement possible from normally stationary objects? Examples are ladders, tools, containers, and supplies.

The Contact-by and Contact-with Types of Accidents

The subtle difference between contact-with and contact-by injuries is that in the first, the agent moves to the victim, while in the second, the victim moves to the agent.

Can the worker be contacted by anything while doing the job step? The contact-by accident is one in which the worker could be contacted by some object or agent. This object or agent is capable of injuring others by nonforceful contact. Examples of items capable of causing injury are chemicals, hot solutions, fire, electrical flashes, and steam.

Can the worker come in contact with some agent that will injure him without forceful contact? Any type of work that involves materials or equipment that may be harmful without forceful contact is a source of contact-with accidents. There are two kinds of work situations which account for most of the contact-with accidents. One situation is working on or near electrically charged equipment, and the other is working with chemicals or handling chemical containers.

The Caught-in and Caught-on Types of Accidents

The next three accident types involve "caught" accidents. Can the person be caught in, caught on, or part between objects? A caught-in accident is one in which the person, or some part of his or her body, is caught in an enclosure or opening of some kind. Can the worker be caught on anything while doing the job step? Most caught-on accidents involve the worker's clothing being caught on some projection of a moving object. This moving object pulls the worker into an injury contact. Or, the worker may be caught on a stationary protruding object, causing a fall.

The Caught-between Type of Accidents

Can the worker be caught between any objects while doing the job step? Caught-between accidents involve having a part of the body caught between something moving and something stationary, or between two moving objects. Always look for the pinch point.

The Fall-to-Same-level and Fall-to-Below Types of Accidents

Slip, trip, and fall accident types are some of the most common accidents occurring in the workplace. Can the worker fall while doing a job step? Falls are such frequent accidents that we need to look thoroughly for slip, trip, and fall hazards. Consider whether the worker can fall from something above ground level, or whether the worker can fall to the same level.

Two hazards account for most fall-to-same level accidents: slipping hazards and tripping hazards. The fall-to-below accidents occur in situations where employees work above ground or above the floor level, and the results are usually more severe.

Overexertion and Exposure Types of Accidents

The next two accident types are overexertion and exposure. Can the worker

→ Hazard Identification

be injured by overexertion; that is, can he or she be injured while lifting, pulling or pushing? Can awkward body positioning while doing a job step cause a sprain or strain? Can the repetitive nature of a task cause an injury to the body? An example of this is excessive flexing of the wrist, which can cause a carpal tunnel syndrome (which is abnormal pressure on the tendons and nerves in the wrist).

Finally, can exposure to the work environment cause an injury to the worker? Environmental conditions such as noises, extreme temperatures, poor air, toxic gases and chemicals, or harmful fumes from work operations should also be listed as hazards.

2. A Hazard Reporting System

Hazard identification is a technique used to examine the workplace for hazards with the potential to cause accidents. Hazard identification, as envisioned in this section, is a worker-oriented process. The workers are trained in hazard identification and asked to recognize and report hazards for evaluation and assessment. Management is not as close to the actual work being performed as are those performing the work. Even supervisors can use extra pairs of eyes looking for areas of concern.

Workers have already hazard concerns and have often devised ways to mitigate the hazard, thus preventing injuries and accidents. This type of information is invaluable when removing and reducing workplace hazards.

This approach to hazard identification does not require that someone with special training conduct it. It can usually be accomplished by the use of a short fill-in-the-blank questionnaire. This hazard identification technique works well where management is open and genuinely concerned about the safety and healthy of its workforce. The most time-consuming portion of this process is analyzing the assessment and response regarding potential hazards identified. Empowering workers to identify hazards, make recommendations on the abatement of hazards, and then suggest how management can respond to these potential hazards is essential.

Unit Six

Words and Expressions

perceive [pə'siːv] *v.* 意识到；发觉；理解
overwhelming [ˌəʊvə'welmɪŋ] *adj.* 势不可挡的；巨大的
budgetary ['bʌdʒɪteri] *adj.* 预算的
constraint [kən'streɪnt] *n.* 约束
audit ['ɔːdɪt] *n.* 审计；检查；*v.* 审计；旁听
bolt [bəʊlt] *n.* 螺栓 *v.* 用螺栓固定
protrude [prə'truːd] *v.* （使某物）伸出；突出
subtle ['sʌt(ə)l] *adj.* 微妙的；敏感的
stationary ['steɪʃənri] *adj.* 固定的；静止的
pinch [pɪntʃ] *n.* 收缩；压力
awkward ['ɔːkwəd] *adj.* 笨拙的；别扭的
sprain [spreɪn] *n.* 扭伤；扭筋
strain [streɪn] *n.* 过度损伤
flexing ['fleksɪŋ] *n.* 挠曲；可挠性
wrist [rɪst] *n.* 手腕；腕关节
carpal ['kɑːpl] *n.* 腕骨；*adj.* 腕骨的
syndrome ['sɪndrəʊm] *n.* 综合征；典型表现
tendons ['tendənz] *n.* [解] 筋；腱
nerve [nɜːv] *n.* 神经；勇气
fume [fjuːm] *n.* 烟；气体

envisioned [ɪn'vɪʒnd] *v.* 设想；想象
questionnaire [ˌkwestʃə'neə(r)] *n.* 调查表；调查问卷
genuinely ['dʒenjuɪnli] *adv.* 真正地；真诚地
time-consuming *adj.* 费时的；旷日持久的
empower [ɪm'paʊə(r)] *v.* 授权；准许
hazard identification 危险因素辨识
chemical process 化工过程
ventilation system 通风系统
expected benefit 期望效益
occupational injury 职业伤害
workplace ['wɜːkpleɪs] *n.* 工作场所；车间
accident investigation 事故调查
preventive control 预防性控制
lead to 导致
safety procedure 安全规程
safety engineering 安全工程
hazard control 危险控制
carpal tunnel syndrome 腕管综合征
abnormal pressure 异常压力
environmental condition 环境条件

Unit Seven

Accident Investigations

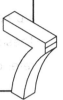

Although accident investigation is an after-the-fact approach to hazard identification, it is still an important part of this process. At times hazards exist, which no one seems to recognize until they result in an accident or incident. In complicated accidents it may take an investigation to actually determine the cause of the accident. This is especially true in cases where death results and few or no witness exist. An accident investigation is a fact-finding process and not a fault-finding process with the purpose of affixing blame. The end of any result of an accident investigation should be to assure that the type of hazard or accident does not exist or occur in the future.

Your company should have a formalized accident investigation procedure, which is followed by everyone. It should be spelled out in writing and end with a written report used as a foundation for your standard company accident investigation form. It may be your workers' compensation form or an equivalent from your insurance carrier.

Accidents and even near misses should be investigated by your company if you are intent on identifying and preventing hazards in your workplace. Thousands of accidents occur throughout the United States every day. The failure of people, equipment, supplies, or surroundings to behave or react as expected causes most of the accidents. Accident investigations determine how and why these failures occur. By using the information gained through an investigation, a similar or perhaps more disastrous accident may be prevented. Accident investigations should be conducted with accident prevention in mind. Investigations are NOT to place blame.

An accident is any unplanned event that results in a personal injury or property damage. When the personal injury requires little or no treatment, it is minor. If it results in a fatality or a permanent total, permanent partial or temporary total (lost-time) disability, it is serious. Similarly, property damage may be minor or serious. Investigate all accidents regardless of the extent of the injury or damage. Accidents are part of a broad group of events that adversely affect the completion of a task. These events are incidents. For simplicity, the procedures discussed in later sections refer only to accidents. They are, however, also applicable to incidents.

1. Accident Prevention

Accidents are usually complex. An accident may have 10 or more events that can be caused. A detailed analysis of an accident will normally reveal three cause levels: basic, indirect, and direct. At the lowest level, an accident occurs only when a person or object receives an amount of energy or a hazardous material that cannot be absorbed safely. This energy or hazardous material is the DIRECT CAUSE of the accident. The direct cause is usually the result of one or more unsafe acts or unsafe conditions, or both. Unsafe acts and conditions are the indirect causes or symptoms. In turn, indirect causes are usually traceable to poor management policies and decisions, or to personal or environmental factors. These are the basic causes.

In spite of their complexity, most accidents are preventable by eliminating one or more causes. Accident investigations determine not only what happened, but also how and why. The information gained from these investigations can prevent the recurrence of similar or perhaps more disastrous accidents. Accident investigators are interested in each event as well as in the sequence of events that led to an accident. The accident type is also important to the investigator. The recurrence of accidents of a particular type or those with common causes shows areas needing special accident prevention emphasis.

2. Investigative Procedures

The actual procedures used in a particular investigation depend on the nature and results of the accident. The agency having jurisdiction over the location determines the administrative procedures. In general, responsible officials will appoint an individual to be in charge of the investigation. An accident investigator should use most of the following steps:

Define the scope of the investigation.

→ Accident Investigations

Select the investigators. Assign specific tasks to each (preferably in writing).

Present a preliminary briefing to the investigating team.

Visit and inspect the accident site to get updated information.

Interview each victim and witness. Also interview those who were present before the accident and those who arrived at the site shortly after the accident. Keep accurate records of each interview. Use a tape recorder if desired and if approved.

Determine the following:

What was not normal before the accident.

Where the abnormality occurred.

When it was first noted.

How it occurred.

Determine the following:

Why the accident occurred.

A likely sequence of events and probable causes (direct, indirect, basic).

Alternative sequences.

Determine the most likely sequence of events and the most probable causes.

Conduct a post-investigation briefing.

Prepare a summary report including the recommended actions to prevent the recurrence. Distribute the report according to applicable instructions.

An investigation is not complete until all data are analyzed and a final report is completed. In practice, the investigative work, data analysis, and report preparation proceed simultaneously over much of the time spent on the investigation.

3. Fact Finding

The investigator collects evidence from many sources during an investigation, gets information from witnesses and observation as well as by reports, interviews witnesses as soon as possible after an accident, inspects the accident site before any changes occur, takes photographs and makes sketches of the accident scene, records all pertinent data on map, and gets copies of all reports. Documents containing normal operating procedures flow diagrams, maintenance charts or reports of difficulties or abnormalities are particularly useful. Keep complete and accurate notes in a bound notebook. Record pre-accident conditions, the accident

·47·

sequence and post-accident conditions. In addition, document the location of victims, witnesses, machinery, energy sources, and hazardous materials.

In some investigations, a particular physical or chemical law, principle, or property may explain a sequence of events. Include laws in the notes taken during the investigation or in the later analysis of data. In addition, gather data during the investigation that may lend itself to analysis by these laws, principles, or properties. An appendix in the final report can include an extended discussion.

4. Interviews

In general, experienced personnel should conduct interviews. If possible, the team assigned to this task should include an individual with a legal background. After interviewing all witnesses, the team should analyze each witness' statement. They may wish to re-interview one or more witnesses to confirm or clarify key points. While there may be inconsistencies in witnesses' statements, investigators should assemble the available testimony into a logical order. Analyze this information along with data from the accident site.

Not all people react in the same manner to a particular stimulus. For example, a witness within close proximity to the accident may have an entirely different story from one who saw it at a distance. Some witnesses may also change their stories after they have discussed it with others. The reason for the change may be additional clues.

A witness who has had a traumatic experience may not be able to recall the details of the accident. A witness who has a vested interest in the results of the investigation may offer biased testimony. Finally, eyesight, hearing, the reaction time, and the general condition of each witness may affect his or her powers of observation. A witness may omit entire sequences because of a failure to observe them or because their importance was not realized.

5. A Report of Investigation

As noted earlier, an accident investigation is not complete until a report is prepared and submitted to proper authorities. Special report forms are available is many cases. Other instances may require a more extender report. Such reports are often very elaborate and may include a cover page, a title page, an abstract, a table of contents, a commentary or narrative discussion of probable causes, and a section on conclusions and recommendations

Accident investigation should be an integral part of your written safety and health program. It should be a formal procedure. A successful accident investigation determines not only what happened, but also finds how and why the accident occurred. Investigations are an effort to prevent a similar or perhaps more disastrous sequence of events. You can then use the resulting information and recommendations to prevent future accidents.

Keeping records is also very important to recognizing and reducing hazards. A review of accidents and injury records over a period of time can help pinpoint the cause of some accidents. If a certain worker shows up several times on the record as being injured, it may indicate that the person is physically unsuited for the job, is not properly trained, or needs better supervision. If one or two occupations experience a high percentage of the accidents in a workplace, they should be carefully analyzed, and countermeasures should be taken to eliminate the causes. If there are multiple accidents involving one machine or process, it is possible that work procedures must be changed or that maintenance is needed. Records that show many accidents during a short period time would suggest an environmental problem.

Once the hazards have been identified, then the information and sources must be analyzed to determine their origin and the potential to remove or mitigate their effects upon the workplace. The analysis of hazards forces us to take a serious look at them.

Words and Expressions

affix [əˈfɪks] v. 归咎于；归于
spell out 使十分清楚明白
compensation [ˌkɒmpenˈseɪʃ(ə)n] n. 补偿；赔偿
disastrous [dɪˈzɑːstrəs] adj. 损失惨重的
fatality [fəˈtæləti] n. 不幸；灾祸
applicable [əˈplɪkəb(ə)l] adj. 可适用的；可应用的
symptom [ˈsɪmptəm] n. 症状；征兆
traceable [ˈtreɪsəb(ə)l] adj. 可追踪的；起源于……
recurrence [rɪˈkʌrəns] n. 重现；循环
jurisdiction [ˌdʒʊərɪsˈdɪkʃn] n. 权限
scope [skəʊp] n. (活动)范围
preferably [ˈprefrəbli] adv. 更适宜
briefing [ˈbriːfɪŋ] n. 简报
approved [əˈpruːvd] adj. 经核准的；被认可的

abnormality [ˌæbnɔːˈmæləti] n. 变态；畸形；异常性
simultaneously [ˌsɪm(ə)lˈteɪnɪəsli] adv. 同时地
sketch [sketʃ] n. 草图；略图
pertinent [ˈpɜːtɪnənt] adj. 有关的；相关的；中肯的
document [ˈdɒkjumənt] v. 证明
inconsistency [ˌɪnkənˈsɪstənsi] n. 矛盾
testimony [ˈtestɪməni] n. 陈述证词
proximity [prɒkˈsɪməti] n. 接近；临近
traumatic [trɔːˈmætɪk] adj. 外伤的；创伤的
vested [ˈvestɪd] adj. 既定的
biased [ˈbaɪəst] adj. (统计试验中)结果偏倚的
pinpoint [ˈpɪnpɔɪnt] n. 精确；adj. 极微小的；v. 查明
mitigate [ˈmɪtɪgeɪt] v. 减轻

Unit Eight
Accident Analysis in Mine Industry

Coal is produced from underground mines in about 50 countries. Underground coal mines range from modern mines using the latest remote-controlled equipment operated by a small, highly skilled workforce benefiting from continuous monitoring of all aspects of workplace conditions, to mines that are dug by hand and where the coal is extracted and transported by hand, often in conditions that are unsafe and unhealthy.

Underground coal mining has historically been one of the highest risk activities as far as the safety and health of workforce are concerned. Fortunately, significant, sustained improvements in coal mining occupational safety and health have been achieved as a result of new technologies, massive capital investment, intensive and continuous training and changes in attitudes to safety and health among those in all stages of the coal chain. Nonetheless, if a safety net, which includes a number of critical checks and balances, is not in place to asses and control the hazards, accidents, ill health and diseases can and do occur. These are discussed as follows.

1. Rock Falls

Underground coal mines frequently suffer from roof falls, which have various consequences ranging from fatalities and injuries to downtime. Underground mining still has one of the highest fatal injury rates of any U.S. industry—more than five times the national average compared to other industries. Between 1996 and 1998, nearly half of all underground fatalities were attributed to roof, rib and face falls. Small pieces of rock falling between bolts injure 500~600 coal miners each year.

Several factors have contribution to occurrences of roof falls in underground coal mines, such as geological and stress conditions, mine layout, mine environment. Among the factors affecting the roof fall hazards in coal mines, stress conditions and mine layout are somewhat controllable by appropriate mine design. However, it is relatively more difficult to control the effect of geological conditions on roof falls, since the geological conditions are the nature's uncertainty, and hence they comprise inherent variability in roof fall occurrence. Therefore, in order to deal with the uncertainties associated with the roof falls, risk assessment methods are required for decreasing the consequences and related costs of roof fall hazards.

2. Outburst

Natural phenomena of sudden gas and rock mass outbursts, such as volcano eruption, geysers, huge bursts of water saturated with CO_2, out of the reservoirs in former volcano craters have been known for a long time. As mining activities upset the natural balance in the rock mass containing the substances that undergo phase transitions, the outbursts of rock and gas may occur. Their occurrences in mines have been recorded for more than 150 years. Attempts have been made to provide an adequate explanation of these processes. Increasing frequency of outburst occurrence after World War II called for still more extensive research.

Coal and gas outburst problems have been exacerbated significantly over the past decade because of higher productivity and the trend towards the recovery of deeper coal seams. However, despite of great efforts, surprisingly little progress has been made towards understanding the outburst mechanism. Prediction techniques continue to be unreliable and unexpected outburst incidents are still a major concern for underground coal mining.

In China outbursts occur in a number of coal fields and in a large number of mines. In fact, the largest number of outbursts have been recorded in China. The most important coal fields where outbursts occur are in the provinces of Shanxi (Yangquan); Liaoning (Beipiao); Henan (Jiaozuo), Chongqing (Nantong and Songzao) and Hebei (Kailuan). Coal and gas bursts are differentiated in China into four categories:

Coal bursts with no gas;
Gas bursts;
Coal and gas outbursts;
Rock and gas outbursts.

3. Mine Fires

Three ingredients are necessary for a fire. These are fuel, oxygen and ignition, referred to as the fire triangle. Coal seams make up a third of the fire triangle with natural deposits of both solid and gaseous fuels. Mine ventilation carries oxygen, the second part of the fire triangle, throughout the mine. Electrical machines, equipment, lights, power stations and circuitry, along with diesel equipment, conveyor belting frictional sources, welding, acetylene cutting and other produces of friction, spark or flame used throughout a mine are ignition sources which add the third ingredient of the fire triangle. To prevent the outbreak of coal mine fires, a number of critical safeguards, checks and balances are necessary.

Fires are a significant hazard to the safety and health of mine workers. Fires at underground and surface mines place the lives and livelihood of miners at risk. Ventilation streams in underground mines can carry smoke and toxic combustion products throughout the mine, making escape through miles of confined passageways difficult and time consuming. A fire in an underground coal mine is especially hazardous due to the ultimate fuel supply and the presence of flammable methane gas. The greatest mine fire disaster in the US occurred at the Cherry Coal Mine, IL, in November 1909, where 259 miners perished. During 1990—2001, more than 975 reportable fires resulted in the temporary closing of several mines. Over 95 of the fires occurred in underground coal mines. The leading causes of mine fires include flame cutting and welding operations, friction, electrical shorts, mobile equipment malfunctions, and welding spontaneous combustion. The prevention, early and reliable detection, control, and suppression of mine fires are critical elements in safeguarding the lives and livelihood of over 230,000 mine workers.

4. Explosions

While much progress has been made in preventing explosion disasters in coal mines, explosions still occur, sometimes producing multiple fatalities. Explosions and the resulting fires often kill or trap workers, block avenues of escape, and rapidly generate asphyxiating gases, threatening every worker underground. Explosions in underground mines and surface processing facilities are caused by accumulations of flammable gas and/or combustible dust mixed with air in the presence of an ignition source. Explosions can be prevented by minimizing methane concentrations through methane drainage and ventilation, by adding sufficient rock dust to inert the coal dust, and by eliminating ignition sources. Explosion effects can be mitigated by using barriers to suppress propagating explosions.

5. Coal Dusts

The production, transportation and processing of coal generates small particles of coal dust. If uncontrolled and allowed to accumulate, that highly explosive dust can be ignited. If it becomes airborne the coal dust can cause violent explosions. Coal dust explosions can create deadly forces, fire and super-heated air which can quickly spread through a mine, killing or injuring several miners. Explosion forces can destroy ventilation and roof controls, block escape routes and trap miners in conditions where oxygen in the mine air is replaced with asphyxiating gases.

The production, transportation and processing of coal generates tiny respirable coal dust particles that become airborne and are invisible to the naked eye. Appropriate instrumentation should be used to quantify the level and size of dust particles present in the air. Coal is made up of a variety of elements. It is mixed with other dusts, most notably crystalline silica, generated from fractured rock in the mine roof, floor or the coal seam which can also become airborne. So coal mine dust can be a significant health risk. When inhaled by miners, dust can result in diseases of the pulmonary system (lungs), including coal workers' pneumoconiosis (CWP), progressive massive fibrosis (PMF), silicosis, and chronic obstructive pulmonary disease (COPD). These lung diseases are progressive, disabling and can be fatal.

6. Electricity

The use of electricity and energized equipment in underground coal mines can result in injuries and deaths from electrical shock or arc burn. Given the confined space of underground mines, which are a dark, and at times a harsh environment, with several pieces of energized equipment and circuitry in close proximity to workers and with self-propelled equipment in motion, the potential of shock or electrocution exists.

Coal mines contain natural deposits of coal, coal mine dust and mine gases that are flammable and explosive. The introduction of electrical and energized equipment in coal mines creates the potential of igniting mine fires and explosions, which can cause numerous deaths and injuries from single events and devastate the mine.

Electrical accidents are the 4th leading cause of death in mining and are disproportionately fatal compared with most other types of mining accidents. About one-fifth of these deaths result when high-reaching mobile equipment contacts power lines overhead. One-half of all mine electrical injuries and fatalities occur during electrical maintenance work, with the following electrical components most commonly involved: circuit breakers, conductors, batteries, and meters.

7. Inrushes of Water, Gas or Other Material

Inrushes of water, noxious or flammable gas or other materials are a serious hazard in coal mining. Mining operations can get too close to old workings or geological abnormalities that contain water, gases or materials that could inundate the mine. One particular hazard is mining next to old workings that were poorly surveyed, not surveyed at all, or not adequately inspected, which contain bodies of water or dangerous mine gases. Old workings filled with water, particularly at elevations higher than the active mine, could quickly flood the mine and drown miners before they could escape if inadvertently cut into. Inrushing mine gases inadvertently encountered can overpower mine ventilation and the oxygen in the air and suffocate miners or, with the right mixture of oxygen, trigger explosions.

Words and Expressions — Unit Eight

extract [ˈekstrækt] n./v. 开采
rock fall 岩石冒落
downtime [ˈdaʊntaɪm] n. 停工
rib [rɪb] n. 矿柱
face [feɪs] n. 采煤工作面
bolt [bəʊlt] n. 锚杆
stress [stres] n. 应力
layout [ˈleɪaʊt] n. 方案；布局
outburst [ˈaʊtbɜːst] n. 突出
geyser [ˈɡiːzə(r)] n. 间隙泉
crater [ˈkreɪtə(r)] n. 火山口
phase transition 相变
exacerbate [ɪɡˈzæsəbeɪt] v. 加剧
mechanism [ˈmekənɪzəm] n. 机理
coal field 煤田
mine fire 矿井火灾
acetylene [əˈsetəliːn] n. 乙炔气
flammable [ˈflæməb(ə)l] adj. 易燃的
methane [ˈmiːθeɪn] n. 甲烷
spontaneous combustion 自然发火
asphyxiate [əsˈfɪksieɪt] v. 使……窒息
concentration [ˌkɒns(ə)nˈtreɪʃ(ə)n] n. 浓度

drainage [ˈdreɪnɪdʒ] n. 抽放
inert [ɪˈnɜːt] adj. 惰性的
barrier [ˈbæriə(r)] n. 隔爆物
crystalline silica 结晶二氧化硅
pulmonary [ˈpʌlmənəri] adj. 肺部的
pneumoconiosis [ˌnjuːm(ʊ)kəʊnɪˈəʊsɪs] n. 尘肺病
fibrosis [faɪˈbrəʊsɪs] n. 纤维症；纤维化
silicosis [ˌsɪlɪˈkəʊsɪs] n. 硅肺病
chronic [ˈkrɒnɪk] adj. 慢性的；延续很长的
shock [ʃɒk] v. 电击
arc burns 电弧灼伤
inrush [ˈɪnrʌʃ] n. 涌入
noxious [ˈnɒkʃəs] adj. 有害的
old working 老采空区；老窑
abnormalities [ˌæbnɔːˈmælɪtiz] n. 畸形；异常性
inundate [ˈɪnʌndeɪt] v. 淹没
survey [ˈsɜːveɪ] n./v. 勘查
elevation [ˌelɪˈveɪʃ(ə)n] n. 开采水平（指高程）
suffocate [ˈsʌfəkeɪt] v. 使……窒息

Unit Nine

Mine Accident Prevention and Control

1. Methods of Dust Control

The natural atmosphere that we breathe contains not only its gaseous constituents but also large numbers of liquid and solid particles. These are known by the generic name aerosols. They arise from a combination of natural and industrial sources including condensation, smokes, volcanic activity, soils and sands, and microflora. Most of the particles are small enough to be invisible to the naked eye. Dust is the term we use in reference to the solid particles. The physiology of air-breathing creatures has evolved to be able to deal efficiently with most of the aerosols that occur naturally. However, within closed industrial environments, concentrations of airborne particulates may reach levels that exceed the ability of the human respiratory system to expel them in a timely manner. In particular, mineral dusts are formed whenever the rock is broken by impact, abrasion, crushing, cutting, grinding or explosives. The fragments that are formed are usually irregular in shape. The large total surface area of dust particles may render them more active physically, chemically and biologically than the parent material. This has an important bearing on the ability of certain dusts to produce lung diseases.

The three major control methods used to reduce the airborne dust in tunnels and underground mines: ventilation, water and dust collectors.

Ventilation: ventilation air reduces the dust through both dilution and displacement. The dilution mechanism operates when workers are surrounded by a dust cloud and additional air serves to reduce the dust concentration by diluting the cloud. The displacement mechanism operates when workers are upwind of dust sources and the air velocity is high enough to reliably keep the dust downwind.

(1) Dilution ventilation. The basic principle behind dilution ventilation is to provide more air and dilute the dust. Most of the time the dust is reduced roughly in

proportion to the increase in airflow but not always. The cost and technical barriers to increased airflow can be substantial, particularly where air already moves through ventilation ductwork or shafts at a velocity of 3,000 ft/min or more.

(2) Displacement ventilation. The basic principle behind displacement ventilation is to use the airflow in a way that confines the dust source and keeps it away from workers by putting dust downwind of the workers. Every tunnel or mine passage with an airflow direction that puts dust downwind of workers uses displacement ventilation. In mines, continuous miner faces or tunnel boring machines on exhaust ventilation use displacement ventilation. Enclosure of a dust source such as a conveyor belt transfer point along with extraction of dusty air from the enclosure, is another example of displacement ventilation. Displacement ventilation can be hard to implement. However, if done well, it is the most effective dust control technique available, and it is worth considerable effort to get it right. The difficulty is that when workers are near a dust source, say, 10~20 ft from the source, keeping them upwind requires a substantial air velocity typically between 60 ft/min and 150 ft/min. There is not always enough air available to achieve these velocities.

Water sprays: the role of water sprays in mining is a dual one: (1) wetting of the broken material being transported; (2) airborne capture. Of the two, wetting of the broken material is far more effective.

Adequate wetting is extremely important for dust control. The vast majority of dust particles created during breakage are not released into the air, but stay attached to the surface of the broken material. Wetting this broken material ensures that the dust particles stay attached. As a result, adding more water can usually (but not always) be counted on to reduce the dust. For example, coal mine operators have been able to reduce the dust from higher longwall production levels by raising the shearer water flow rate to an average of 100 gal/min. Compared to the amount of coal mined, on a weight basis, this 100 gal/min is equivalent to 1.9% added moisture from the shearer alone. Unfortunately, excessive moisture levels can also result in a host of materials handling problems, operational headaches, and product quality issues, so an upper limit on water use is sometimes reached rather quickly. As a result, an alternative to simply adding more water is to ensure that the broken material is being wetted uniformly.

Under actual mining conditions, the typical water spray operating at 100 psi (pound square inch) and 1—2 gal/min gives no more than 30% airborne capture of the respirable dust. This is not as good as lab tests would lead one to believe. In lab tests sprays were usually confined in a duct, and all of the dust was forced to pass through the spray. However, under actual mining conditions, dust clouds are unconfined. In all sprays, the moving droplets exert drag on the adjacent air; thus, sprays act to move the air. Because of this air entrainment effect, if a spray is aimed at an unconfined dust cloud, it will carry in air that spreads the cloud, thus making capture by the spray less efficient. Attempts to improve the airborne capture efficiency of sprays have not met with practical success. One approach has been to reduce the droplet size, based on the notion that capture by smaller droplets is more efficient.

2. Mine Drainage

Drainage in coal mines is extremely variable. In some mines the drainage is excellent and water causes few problems; in others, lack of drainage results in severe inundations. However, even a small amount of water in low coal can cause miner discomfort and result in reduced productivity.

There are both direct and indirect problems associated with the influx of water into a coal mine; these have effects in as well as out of the mine. Some of the direct effects of water in a mine might include: (1) the blocking of mine entries to the passage of air and haulage by the accumulation of water? (2) interruption of production and damage to the mine and possibly even loss of life due to inrush of water; (3) increased costs caused by the need to remove the water; (4) interference with haulage; especially, water can cause poor traction for rubber-tired equipment and result in track deterioration as it washes away ballast.

To minimize the problems associated with an influx of unwanted water into underground mines, a four-step process of control is suggested: (1) prevention; (2) collection; (3) transportation; (4) disposal. Each of them will be reviewed briefly.

1) Prevention

Each gallon of water that is prevented from entering the mine is one less gallon that will have to be collected, transported and disposed of. Some common sense precautions such as avoiding the siting of shafts, boreholes and other opening in the slow spots on the surface and other measures can prevent surface runoff from entering the mine.

2) Collection

Even with the best mine design and all sorts of precautionary measures, water influx into mines can not be entirely eliminated. The water that accumulates has to be collected before it can be transported and disposed of. In the collection process, it is important that gravity should be used as much as possible to minimize the power required to transport the water.

An interesting collection system for mine water is the use of the diversion tunnel. The concept is to predrain a mining area by locating a drainage tunnel under the seams to be mined. If the water can be drained from an area prior to mining, subsequent mining may be done without any water problems. This method appears very attractive since not only does it preclude the handling of water in mines but also the water will remain pure and uncontaminated.

3) Transportation

Even where gravity has been utilized to the greatest possible degree for collecting mine water, some means of transporting the water out of the mine is still required. Today, pumps are used almost exclusively for this purpose.

4) Disposal

With the high effluent standards that must be maintained for streams today, the disposal of mine water can be a very costly problem. There must be some knowledge of the material to be disposed of if it is to be handled properly.

Oxidation of pyrites in the coal seam and strata overlying and underlying the seam is the initial step in the formation of acid mine water. As oxidation continues, the material disintegrates, exposing new surfaces for further oxidation and acid formation. Thus, time is an important factor: the longer the acid-forming materials are exposed to the atmosphere, the greater the amount of acid that will be formed. Thus, the principal methods for treating mine water are designed to neutralize acidity and remove iron by processes involving the use of lime or limestone and by demineralization.

Mine drainage systems contain such items as sumps, suction pipes, pumps, discharge pipes and appropriate fittings such as valves and ells (Fig.5) to move water from a point in the mine to the surface. Energy is introduced into the system via the pump to overcome the following total dynamic water heads (H). The total dynamic head (H) can be expressed mathematically:

$$H = H_s + H_f + H_{sh} + H_v \tag{9.1}$$

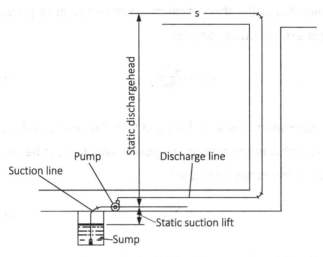

Fig. 5 A mine drainage system

where H_s is the total static head representing difference in elevation between the level of water at the source to the point of discharge, H_f is frictional head produced by resistance of water flowing in the pipes, H_{sh} is shock losses due to changes of water flow produced by fittings, and H_v is the velocity head of liquid moving at a given velocity. It is expressed by the formula

$$H_V = \frac{V^2}{2g} \tag{9.2}$$

where H_v is the velocity head in feet, V is the velocity of ft/s, and g is acceleration due to the gravity ft/s².

Many empirical formulas have been used for calculating the frictional losses in a pipe using the following formula:

$$H_f \propto \frac{fLV^2}{D} \tag{9.3}$$

where *f* is the pipe coefficient of friction, *L* is the length of the pipe, *V* is the velocity of water, and *D* is the diameter of the pipe.

Finally, pressure is required to accelerate the water from rest to the discharge velocity which is generally not recovered. Usually, however, this is so small that it is ignored in the calculations.

Water horsepower (H_p) is the theoretical minimum amount of power required to drive the water and is expressed as follows:

$$H_P = \frac{QH}{3960} \tag{9.4}$$

where *H* is the total dynamic head in feet and *Q* is the flow in gallons per minute.

Brake horsepower (*bhp*) is the amount of power which must be delivered to the input shaft of the pump and is expressed as:

$$H_P = \frac{QH}{3960} \tag{9.5}$$

where *E* is the efficiency of the pump expressed as a decimal and is a ratio of the water horsepower (output) to the brake horsepower (input) times 100.

Words and Expressions — Unit Nine

abrasion [əˈbreɪʒn] *n.* 磨损
respiratory [rəˈspɪrət(ə)ri] *adj.* 呼吸的
aerosol [ˈeərəsɒl] *n.* 浮质
volcanic [vɒlˈkænɪk] *n.* 火山岩；火山的；爆发的
microflora [ˌmaɪkrəʊˈflɒrə] *n.* 微生物群落
physiology [ˌfɪziˈɒlədʒi] *n.* 生理学
all sorts of 一切种类的；各种各样的
aquifer [ˈækwɪfə(r)] *n.* 含水层
ballast [ˈbæləst] *n.* 压块；镇流器；路基；*v.* 使稳定；装重物；铺道楂
be associated with 与……有关；涉及；伴随……；在……的同时
be designed to 设计成；用来；目的是使
brake horsepower 制动功率
common sense 常识
decimal [ˈdesɪm(ə)l] *adj.* 小数的；*n.* 小数
demineralization [diːˌmɪnərəlaɪˈzeɪʃn] *n.* 去矿化（作用）；脱矿质作用
deterioration [dɪˌtɪəriəˈreɪʃ(ə)n] *n.* 恶化；损坏
discharge pipe 排出管；泄水管
discomfort [dɪsˈkʌmfət] *n.* 不安；不适；苦恼时；*v.* 使不安
disintegrate [dɪsˈɪntɪɡreɪt] *v.* 使分离
diversion [daɪˈvɜːʃn] *n.* 转换；导流；分出
effluent [ˈefluənt] *adj.* 流出的；*n.* 流出；废水；废气
ell [el] *n.* L 形短管；弯管

fitting [ˈfɪtɪŋ] *n.* 装配；装配部件；*adj.* 适当的；相称的
flume [fluːm] *n.* 水槽；放水沟；峡沟；*n.* 用水道运输
gallon [ˈɡælən] *n.* 加仑
have some knowledge of 懂得一点
influx [ˈɪnflʌks] *n.* 流入；流注；涌进；汇集；河口
input shaft 输入油
inrush [ˈɪnrʌʃ] *n.* 侵入；开动功率
interference with 干扰
inundation [ˌɪnʌnˈdeɪʃn] *n.* 洪水；淹没；横溢
lack [læk] *v.* 缺乏；*n.* 缺乏；不足
lime [laɪm] *n.* 石灰；氧化钙；*vt.* 用石灰处理；浸在石灰水中
neutralize [ˈnjuːtrəlaɪz] *v.* 使中和；平衡
precautionary [prɪˈkɔːʃənəri] *adj.* 预防的；警戒的
preclude [prɪˈkluːd] *v.* 预防；消除
resurrect [ˌrezəˈrekt] *v.* 复活；使再现
shock loss 局部阻力损失；净击损失
spot [spɒt] *n.* 污点；地点；部位；*v.* 沾上污渍；定点；*adj.* 现场的；现货的
suction pipe 吸水管
traction [ˈtrækʃn] *n.* 牵引力；推力
uncontaminated [ˌʌnkənˈtæmɪneɪtɪd] *adj.* 未污染的；无杂质的；洁净的

·63·

Unit Ten

The System Safety Process

The system safety process is really an easy concept to grasp. The overall purpose is to identify hazards, eliminate or control them, and mitigate the residual risks. The process should combine management oversight and engineering analyses to provide a comprehensive, systematic approach to managing the system risks, and Fig. 6 details this process.

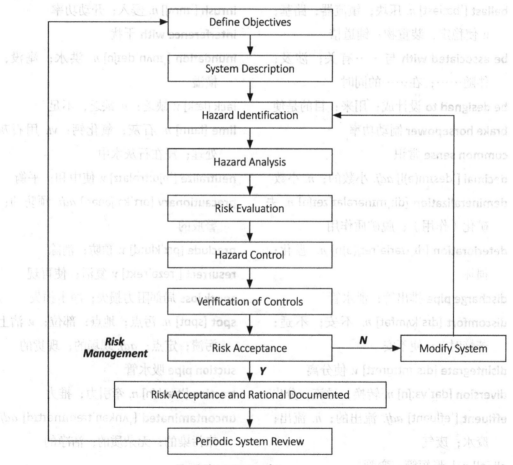

Fig. 6 The system safety process

→ The System Safety Process

As with an problem, the first step is to define the boundary condition or analysis objectives. That is the scope or level of protection desired. One must understand what level of safety is desired at what cost. The engineer needs to answer the question: How safe is safe enough? Other questions to ask are:

What constitutes a catastrophic accident?

What constitutes a critical accident?

Is the cost of preventing the accident acceptable?

Most industries approach this step in the same way. However, how they differentiate among catastrophic, critical, minor, and negligible hazards may vary. The engineer will need to modify the definitions to fit the particular problem. What is important is that these definitions are determined before work begins. A rule-of-thumb definition for each is:

Catastrophic—any event that may cause death or serious personnel injuries, or the loss of system (e.g., an anhydrous ammonia tanker truck overturns, resulting in a major spill).

Critical—any event that may cause severe injuries, or the loss of the mission-critical hardware or high-dollar-value equipment (e.g., the regulator fails open and over-pressurizes a remote hydraulic line, damaging equipment and bringing the system down for some days).

Minor—any event that may cause minor injuries or minor system damages, but does not significantly impact the mission (e.g., the pressure control valve fails open, causing pressure drops and increased caustic levels).

Negligible—any event that does not result in injuries or system damages and does not affect the mission (e.g., lose commercial power, causing the shut down of plant cafeteria).

The next step is system description. Some time should be given to grasp how the system works and how the hardware, software, people, and environment all interact. If the system is not described accurately, then the safety analysis and control program will be flawed.

1. Hazard Identification

Hazard identification is a crucial part of the system safety process. It really is impossible to safeguard a system or control risks adequately without first identifying the hazards. An all-too-frequent mistake in safety engineering is to skip over this step, or not to give it adequate attention. The hazard identification process is a kind of "safety brainstorming". The purpose is to identify hazards as many as are possible and credible. Through this process the engineer develops a preliminary hazard list (PHL) and later will assess the impact on the system.

To develop a PHL the engineer will want to use various methods to gather the most exhaustive list possible. This may refer to:

Survey the site;
Interview site personnel;
Convene a technical experts panel;
Analyze and compare similar systems;
Identify codes, standards, and regulations;
Review relevant technical data (electrical and mechanical drawings, analyses, operator manuals and procedures, engineering reports, etc.);
Analyze energy sources (voltage/current sources, high/low temperature sources, etc.).

The next step is to analyze the hazards identified. A hazard analysis is a technique for studying the cause/consequence relation of the hazard potential in a system. The purpose is to take the preliminary hazard list one level deeper and assess how each hazard affects the system. Is it catastrophic? Or is it critical? The hazard analysis will also assist the engineer in further assessing if further study is needed. There are various hazard analysis techniques that are commonly used in different industries.

After hazards have been identified and analyzed, the engineer needs to control their occurrence or mitigate their effects. This is done by evaluating the risks. Is the hazard likely to occur? The engineer needs to understand the relationship between hazard cause and effect. With this information, the associated risks are then ranked and engineering management is better able to determine which risks are worth controlling and which risks require less attention.

2. Hazard Control

After evaluating the risks and ranking their importance, the engineer must control their effects. Controls fall into two broad categories: engineering controls and management controls. Engineering controls are changes in the hardware that either eliminate the hazards or mitigate their risks. Some example engineering controls include: adding a relief valve to a 2000-psi oxygen system; building a berm around an oil storage tank; using only hermetically sealed switches in an explosive environment; or putting in hard stops in rotating machinery to prevent overtorquing.

Management controls are changes made to the organization itself. Developing and implementing a plant safety plan is a good method of applying management controls to hazards. Some examples are: using production-line employees as safety representatives for their areas; requiring middle-management reviews and approvals of any plant or system modifications to consider safety implications; or assigning signature authority to safety engineers for all engineering change orders and drawings.

Once controls are in place, a method needs to be used to verify that the controls actually control the hazards or mitigate the risks to an acceptable level. Verification of hazard controls is usually accomplished through the company or the engineering management structure. The most frequent means is inspection. However, as we all know, inspection is also one of the most expensive ways to assure that controls are in place. An effective method of hazard control verification is the use of a closed-loop tracking and resolution process.

3. Risk Acceptance

Safety is only as important as management wants to make it. At this point in the safety process this becomes obvious. After the system has been studied and hazards identified, then analyzed and evaluated with controls in place, management must make the formal decisions of which ones they will not take. At this point a good cost-benefit analysis will help management make that decision. Sometimes this is not easy.

Part of the risk acceptance process is a methodical decision-making approach. If the risks are not acceptable, then the system must be modified and the hazard identification process must be followed once again. If the risks are acceptable, then good documentation with written rationale is imperative to protect against liability

10 → → Unit Ten

claims.

Probably one of the key points of the system safety process is that it is a closed loop system. This means that the engineering and management organizations periodically review the safety program, engineering processes, management organizations, and product field use. The American automobile industry has lost billions of dollars in automobile recalls due to safety problems, some of which possibly could have been avoided by a periodic review of product use.

 **Words and Expressions**

 **Unit Ten**

rule-of-thumb 单凭经验的方法
catastrophic [ˌkætəˈstrɒfɪk] *adj.* 灾难的；悲惨的
anhydrous ammonia [ænˈhaɪdrəs əˈməʊniə] 无水氨
hydraulic [haɪˈdrɒlɪk] *adj.* 液压的；水力的；水力学的
caustic [ˈkɔːstɪk] *adj.* 腐蚀性的
flawed [flɔːd] *adj.* 有缺陷的；有瑕疵的；有裂纹的

all-too-frequent 太频繁
code [kəʊd] *n.* 规范；规程
mitigate [ˈmɪtɪɡeɪt] *v.* 使缓和；使减轻
hermetically [hɜːˈmetɪkli] *adv.* 密封地
closed-loop 闭环
verification [ˌverɪfɪˈkeɪʃn] *n.* 确认；查证；核实
rationale [ˌræʃəˈnɑːl] *n.* 基本原理

Unit Eleven
Introduction to the Safety System Engineering

1. Basics of Safety System Engineering
1) System

System is the study object of system engineering, which is a combination of people, procedures, facility, and/or equipment all functioning within a given or specified working environment to accomplish a specific task or a set of tasks. A system embodies four characteristics:

(1) Integrity

A system is a complete whole composed by two distinct elements (components or subsystems). Different elements form an unified whole which is equipped with certain new functions through synthesizing and unifying instead of random assembling although the elements have different functions. Only when new functions appear, may a system perform functions properly and effectively.

(2) Relativity

Elements in a system are interactive and independent, through which the system is linked effectively and may perform its functions.

(3) Purposefulness

Any system aims at certain tasks or goals. In order to fulfill certain aims of a system, prescribed functions must be embodied so as to offer the system optimal planning, designing, controlling, managing and so on in the system life cycle, i. e. the phases of planning, designing, experimenting, manufacturing and applying.

(4) Environment adaptability

Any system is in a certain physical environment, so it must adapt to the changes of the environmental conditions. The effects of environments on the system must be taken into consideration in system research.

2) System Engineering

As a new science in 1950s, system engineering studies system, applying modern science and technology with the aim of system optimization.

System engineering is a scientific methodology of planning, designing, manufacturing, experimenting and applying in organization and management systems, and it is applicable to any systems.

3) SSE

SSE, based on basic theories and methodologies of system engineering, pre-identifies and analyzes hazardous elements existing in a system, assesses and controls them to ensure the expected system safety. To be more specific:

(1) The theoretical basis of SSE is the science of safety and system. It is a system engineering of industrial and mining enterprises with the field of occupational safety and health.

(2) The purpose of SSE is pursuing the safety of the whole process of the system as well as its whole process.

(3) The key point of SSE is the identification, analysis and assessment of system hazardous elements, system safety decision and accident control.

(4) The expected safety target of SSE is to control system risks to an acceptable level. That is to say, to control accidents economically and effectively and keep system risks under the safety index.

2. Research Objects, Contents and Methods of SSE

1) Research Objects of SSE

As a scientific approach, SSE has its own research objects. Any manufacturing system is composed of three factors: first, the operating and managing personnel, who are engaged in manufacturing activities; second, physical conditions, which are necessary to manufacturing such as machines and factory buildings; third, an environment in which manufacturing takes place. The above three factors comprise the "man-machine-environment system"; each factor is one subsystem of the whole system and they are called man subsystem, machine subsystem, and environment subsystem, respectively.

(1) Man subsystem

Whether the man subsystem is safety or not involves the problems as the following, physiological and psychological factors of human, the properly whether the prescriptions, regulation criteria, managing methods, etc. suit human or not and whether it's likely to be accepted by human. During the study, man should be regarded not merely as one in nature or in a company, but as one in a society. Problems must be analyzed and solved from the perspectives of sociology, anthropology, psychology and behavior science. In the man subsystem, man should not only be looked on as an eternal composing part, but also as a kind of creature of self-respect and self-love, with feelings, thoughts and subjective initiatives.

(2) Machine subsystem

For this subsystem, the safety of the workpiece shape, size, material, intensity, technology and reliability of equipments should be taken into account. Meanwhile, requirements of meters and operating pieces on man, and requirements of meters and operating pieces from the perspective of anthropometry, physiology, relevant parameters of psychology and physiology process designing should not be ignored.

(3) Environment subsystem

For this subsystem, physicochemical and social factors need to be considered, physicochemical factors include noise, vibration, dust, poisonous gases, cosmic ray, sunlight, temperature, humidity, pressure, heat, chemical hazardous substance, and so on, and social factors are composed of the management system, time quota, personnel organization, interpersonal relationship, etc.

The three subsystems interact and interplay, which make the safety of the overall system in a certain state. For example, physicochemical factors affect the life cycle and precision of machines or even damage machines; noise, vibration, temperature, dust, and poisonous gases of the machine subsystem affect the man and environment subsystems; the states of human's psychology and physiology are usually subjective factors of mal-operation; social factors in the environment subsystem will affect human's psychology and bring potential hazards to safety. That is to say, the above three subsystems, which interact, restrict and influence each other form an organic whole of the "man-machine-environment". Only starting from the relations in and between these subsystems, can safety problems be really solved in analyzing, assessing and controlling the safety of the "man-machine-environment system". The object of SSE is the "man-machine-environment system".

2) Research Contents of SSE

SSE is a scientific approach (technique) which studies how to utilize theories and methodologies of system engineering to ensure system safety. The main technological methodologies are system safety analysis, system safety prediction, system safety assessment, and system risk control.

(1) System safety analysis

In order to increase system safety and reduce or exterminate accidents, the prerequisite is to identify the hazardous elements in advance, then to master their characteristics thoroughly and to understand clearly the degree of their effects on system safety. Only in this way, can the main potential hazards be found out and effective protection measures be taken to improve system safety. "Preliminary" means that, no matter which stage the system is in its life cycle, safety analysis must be carried out, hazardous factors be identified and mastered before the operation of certain stage. This is what needs to be solved in system safety analysis.

System safety analysis is a kind of technical approach which employs theories and methodologies of system engineering to identify and analyze the existing hazards in a system and make qualitative and quantitative descriptions with reference to practical needs. According to relevant references, there are many forms and methods on system safety analysis and cautions in operation as well.

① Based on the system characteristics, the requirements and aims of analysis, employ different analysis methods for the research in that no method is applicable to any cases because of its characters and limitations. Several methods need to be used simultaneously to make up for limitations of each other or make comparisons; hence, the results analysis can be verified.

② The methods should not be utilized mechanically, they need to be reformed or simplified when necessary.

③ Never be restricted to the application of the methods, but explore new ways and methods based on system theories instead. Meanwhile, the former effective methods also need to be improved to form systematic safety analysis methods.

(2) System safety prediction

Prediction is a science of studying future. With the development of science and technology, and the development of production, human's prediction activities are more and more frequent, prediction fields broader, prediction methods and techniques more advanced. It is widely applied to many fields, such as economy, technology and social development. The effective combination of prediction techniques and controlling methods is the main stream of modern safety technology.

(3) System safety assessment

System safety assessment, based on system safety analysis, identifies the hazardous elements in the system. However, that does not mean that measures must be taken to all the hazardous elements. Instead, it needs to have a good grasp of the accident risks through system safety assessment and then compare them with relevant system safety index. If the accident risks are above the index, measures need to be taken to control the hazards with the aim of keeping them below the index.

(4) System risk control

Any system safety techniques of prediction, analysis or assessment will not be able to take effect as expected if there are no effective management skills and methods. Therefore, system risk control develops with system safety techniques of prediction, analysis and assessment. Its key characteristics is to carry out safety management on the system in full scale during its whole process based on the integration, relevance and orderliness of the system, so as to control system safety.

3) Research Methods of SSE

Based on the theory of safety science, a research method of SSE develops and matures on the basis of former research methods. Generally speaking, there are five methods:

(1) From the integrality of a system

Research methods of SSE must start from the integrality of the system, on which the methods process and targets of solving safety problems are considered. For example, safety requirements on each subsystem must be consistent with that of the whole system and other functions. In the process of system research, contradictions between subsystems and the system or between subsystems themselves must make use of system optimization to find solutions which are acceptable to each part. Meanwhile, system optimization should be utilized through all the phases of system planning, designing, developing and applying.

(2) Essential safety approach

The essential safety approach is the aim of safety technology, and it is also the kernel of SSE approaches. Due to the consideration of the man-machine-environment in safety as one "system", the kernel aim is essential safety, which is the methods and techniques of system essential safety regardless of studying from system contents or objectives.

(3) The match of man and machine

The match of man and machine is a crucial factor among those that affecting system safety. The man-machine match which studies safety in industry is safety man and machine engineering; while in the field of human living, the study of the match between man and machine is the problems of ecological environment and humanistic environment. Obviously, starting from the safety objective, it is an important supporting point of SSE to study the man-machine match and apply its theories and techniques.

(4) Safety economy approach

According to safety relevance, there is correlation between input and objective of safety in certain economic and technological conditions. That is to say, the optimization of system safety is also restricted to economy. However, because of the peculiarity of safety economy (the permeability of safety input and output, the foresight of safety input and the hysteretic nature of safety output, the multi-objective of safety output assessment, the effectiveness of safety economy input and output), SSE should have leading ideas and techniques when studying system targets. Meanwhile, the safety target should be expressed and calculated in many ways.

(5) System safety managing approach

SSE combines technology and management from the perspective of subject. It is a scientific method of solving safety problems with respect to the scientific system theory. SSE is the professional technique basis that integrates the theory and practices. The system safety managing approach permeates through the whole process of safety planning, designing, checking and controlling. Therefore, the approach is a crucial element of SSE.

3. Origin and Development of SSE
1) The Origin of SSE

SSE originates in the 1960s in industrialized countries such as America and Britain. In this period, due to the four disastrous accidents in the research of the American missile system in only one and a half years, American air forces had to study the safety and reliability of the missile system by the basic principles and managing methods of system engineering. USAF proposed "SSE for the Development of Air Force Ballistic Missiles" (in 1962) and issued Weapon System Safety Standards. In 1963, System Safety Program was proposed. In July 1967, it was formalized as criteria of American troops by the U. S. Department of Defense (DOD). After two revisions, it became the nowadays MIL-STD-882B, entitled System Safety Program Requirements, which standardizes safety requirements, managing programs, managing methods and managing objectives in projects bids, the studying and development of the American military system. This is the SSE in the military system brought about by accidents that happened.

In the mid of 1960s, Britain established the system reliability service station and reliability data basic, and successfully developed the probabilistic risk assessment (PRA) technique through which the risks and the acceptability of the nuclear power station system can be calculated. In 1974, American Atomic Energy Commission (AEC) published *The Reactor Safety* (WASH-1400) by professor William G. Johnson, which developed and applied the techniques of system safety analysis and assessment. This report was proved scientific and precise in accident prediction by "Three-Miles Island Incident" (radioactive substance leaking because nuclear piles fused). This is the SSE in the nuclear industry.

The Dow Chemical Company published Dow's Fire & Explosion Index, which is shorted as Dow Method. This method has undergone six revisions in practice, and the seventh edition was published. Based on the substance coefficient which is fixed on the physicochemical characteristics of chemical substances, this method integrates hazardous characteristics of both general and special technological processes, calculation fire and explosion index for system, assesses system losses, and considers safety measures to modify system risk index based on the above. After that, Imperial Chemical Industries Ltd. (ICI) developed the MOND Fire, Explosion and Toxicity Index. In the 1970s, a novel assessing method was published in Labor Province, Japan, Guide to Chemical Industries Safety Assessment, also called Six Phases Safety Analysis. This method is characterized by the combination of analysis and assessment, qualitative and qualitative assessment. It is a managing standard of assessing the whole process of the chemical system. Furthermore, it not only regulates the assessing methods, assessing techniques but regulates which method to employ and how to assess as well.

SSE originates and develops in mass civilian industries. The 1960s saw the American market becoming increasingly competitive. Many new products poured into the market without safety guarantee, which caused many accidents. Consumers required the producers to compensate for their losses. Some even resorted to law demanding producers to be responsible for their criminal acts. Producers were forced to seek new ways and methods to improve products safety as they were developing new products. Many system safety analysis methods and techniques were developed in the fields of electron, aviation, railway, car producing and metallurgy.

2) Development of SSE

Nowadays, SSE has aroused the attention of many countries, and it has gained rapid development. In China, SSE was studied and developed at the end of the 1990s. Oriental Chemical Industry in Tianjin successfully solved the problems of safe products of perilous enterprises employing SSE, which served as a guide to learning and applying SSE in various fields. Afterwards enterprises of different fields borrowed system safety analysis methods from overseas to analyze their own systems. Since the mid and end of the 1980's, people's focus had shifted to assessing theories and methods of system safety and developed many system safety assessment methods, among which the industrial safety analysis method worked out the assessment of industry risk and safety management.

During this period, relevant works written by experts and scholars in the field summarized the theories and techniques on the safety system at home and abroad in a systematic way. Generally, the contents of these works based on the principles

11 → → Unit Eleven

of system engineering, include three parts: system safety analysis, assessment and management. The risk can be identified through analysis and assessment, controlled through program technology management so that system safety can realize the expected goal.

Based on the above, the techniques and methods of system safety analysis, assessment and management published overseas so far belong to SSE; while the techniques and methods of system safety analysis, assessment and management that have been proved effective in controlling preliminary accidents and improving system safety are also crucial parts of SSE.

 **Words and Expressions**

 Unit Eleven

embody [ɪmˈbɒdi] v. 体现；包含
integrity [ɪnˈtegrəti] n. 完整；正直
subsystem [səbˈsɪstəm] n. 子系统；分系统
synthesize [ˈsɪnθəsaɪz] v. 合成；综合
unify [ˈjuːnɪfaɪ] v. 整合；联合
assemble [əˈsemb(ə)l] v. 集合，聚集；装配
relativity [ˌreləˈtɪvəti] n. 相对论；相关性
interactive [ˌɪntərˈæktɪv] adj. 交互式的；相互作用的
purposefulness [ˈpɜːpəsflnəs] n. 目的性；意志坚强
fulfill [fʊlˈfɪl] v. 履行；实现
prescribe [prɪˈskraɪb] v. 规定；开药方
prescription [prɪˈskrɪpʃn] n. 药方；处方；指示
optimization [ˌɒptɪmaɪˈzeɪʃən] n. 最优化；最佳化
methodology [ˌmeθəˈdɒlədʒi] n.（从事某一活动的）一套方法；方法学；方法论
enterprise [ˈentəpraɪz] n. 公司；企业单位
criteria [kraɪˈtɪəriə] n. 标准；条件
anthropology [ˌænθrəˈpɒlədʒi] n. 人类学
vibration [vaɪˈbreɪʃn] n. 摆动；震动
poisonous [ˈpɔɪzənəs] adj. 有毒的；极令人厌恶的
eternal [ɪˈtɜːn(ə)l] adj. 不朽的；永久的
workpiece [ˈwɜːkpiːs] n. 工件；工作部件
anthropometry [ˌænθrəˈpɒmɪtri] n. 人体测量学

cosmic [ˈkɒzmɪk] adj. 宇宙的；外太空的
interplay [ˈɪntəpleɪ] n. 相互作用
precision [prɪˈsɪʒn] n. 精确度；准确（性）
precise [prɪˈsaɪs] adj. 精确的；明确的
mal-operation 错误操作；不正确运行
psychology [saɪˈkɒlədʒi] n. 心理（学）；思想
physiology [ˌfɪziˈɒlədʒi] n. 生理学；生理机能
exterminate [ɪkˈstɜːmɪneɪt] v. 消灭；根除
prerequisite [ˌpriːˈrekwəzɪt] n. 先决条件；前提
preliminary [prɪˈlɪmɪnəri] adj. 初步的；初级的
hazardous [ˈhæzədəs] adj. 冒险的；有危险的；碰运气的
integration [ˌɪntɪˈgreɪʃ(ə)n] n. 集成；结合
orderliness [ˈɔːdəlinəs] n. 整洁；有序性
integrality [ˌɪntəˈgræləti] n. 完整性
kernel [ˈkɜːn(ə)l] n. 核心；要点
crucial [ˈkruːʃ(ə)l] adj. 至关重要的；关键性的
humanistic [ˌhjuːməˈnɪstɪk] adj. 人文主义的
peculiarity [pɪˌkjuːliˈærəti] n. 特性；特质
permeate [ˈpɜːmieɪt] v. 弥漫；遍布；渗入
disastrous [dɪˈzɑːstrəs] adj. 灾难性的；极糟糕的
missile [ˈmɪsaɪl] n. 导弹；投射物
formalize [ˈfɔːməlaɪz] v. 使正式；形式化
revision [rɪˈvɪʒn] n. 修订；修改

 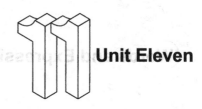

Words and Expressions — Unit Eleven

probabilistic [ˌprɒbəbəˈlɪstɪk] *adj.* 概率性的；或然说的
toxicity [tɒkˈsɪsəti] *n.* 毒性
criminal [ˈkrɪmɪn(ə)l] *adj.* 犯罪的；刑法的；*n.* 罪犯
aviation [ˌeɪviˈeɪʃn] *n.* 航空；飞机制造业
metallurgy [məˈtælədʒi] *n.* 冶金；冶金学；冶金术
perilous [ˈperələs] *adj.* 危险的；冒险的
study object 研究对象
a given or specified working environment 给定或指定的工作环境
complete whole 完整整体
unified whole 统一整体
aim at 针对；瞄准；目的在于
so as to 以便；为了
in system life cycle 在系统生命周期中
taken into consideration 考虑到
key point 关键点；要点
engaged in 忙于；着手于
man subsystem 人子系统
subjective initiative 主观能动性
taken into account 考虑；重视
time quota 时间配额
qualitative and quantitative 定性和定量
make up for 补偿；弥补
have a good grasp of 深刻了解；很好掌握
accident risks 事故风险
take effect 生效；奏效
in full scale 全面
be consistent with 与……一致
make use of 使用；利用
safety objective 安全目标
safety input and output 安全投入和产出
hysteretic nature 滞后性
with respect to 关于；至于
integrates theory and practice 理论与实践相结合
one and a half years 一年半
substance coefficient 物质系数
hazardous characteristics 危险特性
resorted to 诉诸
home and abroad 国内外；海内外
so far 到目前为止；迄今为止

Unit Twelve

Introduction to Fault Tree Analysis

Fault Tree Analysis (FTA) is a common analytical method in SSE. In 1961, H. A. Watson of American Bell Labs proposed FTA and applied it to safety assessment of Minuteman Guidance System to predict the stochastic failure probability of missile launch. There after, Hassle in Boeing Company made great improvements and applied computers to assist the analysis and calculation. In 1974, Atomic Energy Commission executed risk assessment on the commercial nuclear power station by means of FTA and published *Rasmussen Report*, which drew the attention of the world. At the moment, FTA has been applied in the fields of electron, electric power, chemical industry, machinery industry, transportation from the aircraft industry and the nuclear power industry. It can diagnose failures, analyze system weak components, guide system safe operation and maintenance and realize system optimization design.

1. Definition of FTA

FTA is a deductive reasoning, which can signify the logic relations between system possible faults and their causes by means of FT. Through qualitative and quantitative analysis of FT, major causes of faults are identified, which will offer a solid foundation to safety countermeasures to predicate and prevent faults.

2. Procedures of FTA

FTA, based on the information of possible faults or former faults in a system, investigates relevant causes of the fault and takes effective safeguards to prevent accidents. The following are basic procedures of FTA and analysts may choose several ones according to analytical needs and practical conditions when executing analysis on a certain system.

1) Seedtime

(1) Define the system. Deal with the relations among the target system, its external environment and boundary conditions properly, identify the system boundary, and define the major factors affecting system safety.

(2) Get familiar with the system. This is the basis and foundation of FTA. Collect relevant materials and data including the system structure, performance, technological process, operation condition, fault types, maintenance instances and environment factors when doing research on the definite system.

(3) Investigate faults of the system. Collect and investigate the former faults and possible faults in the future. At the same time, faults of the same system in other industries both at home and abroad as well need collecting and investigating.

2) Construct FT

(1) Define top Undesired Event (UE) of Fault Tree (FT). To define the top UE means determining the object event which needs analyzing. Assess its loss and faults frequency based on the faults investigation report. Choose those that are prone to happen and those that cause severe results as the top UE.

(2) Investigate all the reason events relevant to the top UE. Investigate all the reason events relevant to the top UE in terms of human, machine, environment and information so as to identify fault causes and carry out effect analysis.

(3) Construct FT. Arrange the top UE and the reason events which cause the top UE applying definite symbols according to their logic to compile a tree diagram which reflects the cause and result relation.

3) Qualitative assessment of FT

Qualitative assessment calculates the minimal Cut Set (CS) or minimal Path Set (PS) and the structural importance of basic events based on the FT structure. Safeguards of preventing faults are determined according to the results of qualitative assessment.

4) Quantitative assessment of FT

The probability of the top UE is worked out based on the probability of each basic event which causes the fault. The probability importance degree and criticality importance degree of each basic event are worked out as well. Risk assessment of

→ Introduction to Fault Tree Analysis

the system is executed according to quantitative assessment results and possible harms after the fault.

5) Summary and application of FTA results

A timely assessment and a summary of FTA results must be made to propose improvement suggestions. All the materials and data of FTA qualitative and quantitative assessment need to be organized and stored. Safety assessment materials also need to be used comprehensively so as to provide a foundation for system safety assessment and safety design.

3. Symbols and Means of FT

The symbols of FT are divided into three categories: event, logic gate and transfer symbols.

1) Event and symbols

In FTA, the abnormal state or condition is an undesired event; while the sound state or normal condition is a success event. They are both events. Every node in FT is an event.

(1) Resultant event. A resultant event is caused by other events or combinations of events. It is always at the output terminal of a logic gate. It is represented by a rectangle as in Fig.7 (a). Result events are classified into the top event and the intermediate event.

① A top event is the resultant event which is concerned in FTA. It is at the top of FT, which always concerns about the output event instead of the input event of logic gates in FT. That is to say, it discusses possible or the existing fault results in a system.

② An intermediate event is the resultant event which lies between the top event and the basic event in FT. It is the out put event of a logic gate and the input event of other logic gates as well.

(2) Basic event, A basic event is a reason event which causes other events and lies at the root of FT. It is always the input event of a logic gate but never an output event. The basic event is further divided into the basic event and the ellipsis event.

① A basic event is the most fundamental reason or defect event which causes the top event or cannot be further analyzed. It is represented by a circle as is in Fig.7 (b).

② An ellipsis event is a reason event that requires no further analysis or which does not have clear reasons. In addition, an ellipsis event can also be used to refer to a secondary failure, i. e. it is not the reason event of its own system but that from outside the system. An ellipsis event is represented by a diamond as in Fig.7(c).

(3) Special event. A special event is one which needs attention or that requires its peculiarity highlighted. Special events have two categories: the switch event and the conditional event.

① A switch event is also called a normal event, which either occurs or does not occur during normal system operation. See the house in Fig.7(d).

② A conditional event is the event which inhabits the initiation of logic gates. See the oval in Fig.7 (e).

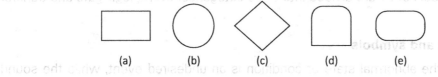

Fig. 7 Symbol of FT

2) Logic gates and symbols

Logic gates connect events and show their logic relations.

(1) AND gate. An AND gate can connect several input events, such as $E_1, E_2, ..., E_n$, and one output event E. It refers to the logic relation in which the output event E will occur only if all of the input event exist simultaneously. Its symbol is shown in Fig.8 (a).

(2) OR gate. An OR gate can connect several input events, such as $E_1, E_2, ..., E_n$, and one output event E. The output event E will occur if any of the input events occurs. Its symbol is shown in Fig.8 (b).

(3) NOT gate. A NOT gate indicates that the output event is opposite to the input event. The symbol is shown in Fig.8 (c).

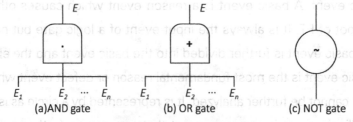

Fig. 8 Logic gates and symbols

(4) Special gates

① Voting gate. Only when m (m/n) or more than m input events happen, the output event occurs. The voting gate symbol is shown in Fig.9 (a). It is obviously that the OR gate and the AND gate are special cases of the voting gate. If m = the voting gates, it is an OR gate; if $m = n$, the voting gate is an AND gate.

② XOR gate. The output event occurs if only exactly one input event occurs. Its symbol is shown in Fig.9 (b).

③ Inhibit gate. The output event occurs if the input event occurs and the conditional event occurs. Its symbol is shown in Fig.9 (c).

④ Exclusive OR gate. The output event occurs if the input event occurs simultaneously and condition A is met. Its symbol is shown in Fig.9 (d).

⑤ Priority AND gate. The output event occurs if at least one input event occurs and condition A is met. Its symbol is shown in Fig.9 (e).

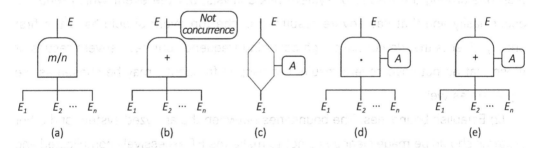

Fig. 9 Special gats and symbols

(5) Transfer symbols

It contrains many same parts, as shown in Fig.10.

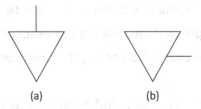

Fig. 10 Transfer symbols

4. Regulations of Construct FT

Construct FT is the most fundamental and critical link in FTA. It is completed by a team composed of system designers, operators and reliable analysts. The FT must undergo repeated research and further development so as to approach perfection. During the process team members will have a further understanding of the system and identify weak parts, which is the most primary purpose of compiling FT. The perfection of FT has direct effects on the correctness of qualitative and quantitative analytical results, and the success of FTA. Therefore, the timely summary of experience in the compiling practice is rather important.

The compiling methods are commonly divided into two categories: artificial construct and computer assistant construct.

Constructing a FT is a rigorous logic reasoning process. It should follow the following regulations:

(1) Undesired events of high risks ought to be preferentially considered in identifying the top event. Whether the right top event is chosen has direct connection with analytical results, so it is the key of FT A. The undesired events are far more than one among the results of system risk analysis, but the event which tends to occur easily and that has severe results, i. e. the top event should have the first priority. Events that do not have high occurring frequency but have severe results or those that do not have severe results but happen frequently may be chosen as the top event as well.

(2) Establish boundaries. The boundaries between the analyzed system and other systems should be made clear so as not to make the FT excessively complicated and confusing after the top event is determined. Meanwhile, some necessary and proper hypotheses can be made as well.

(3) Keep the integrity of gates. No gate-to-gate connections are allowed (always have a text box). FT should be constructed step by step and any leap is not allowed. Any logic gate output must have a result event; gates are not to be directly connected leaping the result event. Otherwise, the precision of logic relation will not have a solid foundation.

(4) Describe the top event precisely. Define the top event clearly, that is to say, describe the accident exactly, when and how it occurred.

(5) Make proper and timely simplification in and after the construct.

 **Words and Expressions**

 **Unit Twelve**

stochastic [stɒˈkæstɪk] *adj.* 随机的
diagnose [ˈdaɪəgnəʊz] *v.* 诊断；判断
optimization [ˌɒptɪmaɪˈzeɪʃən] *n.* 最佳化；最优化
deductive reasoning 演绎推理
countermeasure [ˈkaʊntəmeʒə(r)] *n.* 对策；反措施
investigate [ɪnˈvestɪgeɪt] *v.* 调查；研究
boundary [ˈbaʊndri] *n.* 边界；分界线
define [dɪˈfaɪn] *v.* 给……定义；阐明；限定
material [məˈtɪəriəl] *n.* 材料；原料
maintenance instance 维修实例
prone to 易于；倾向于
Cut Set 割集
Path Set 径集
safeguard [ˈseɪfgɑːd] *n.* 保护；保卫；安全措施；*v.* 防护；保卫
probability importance degree 概率重要度
criticality importance degree 临界重要度
risk assessment 风险评估
propose [prəˈpəʊz] *v.* 建议；*vi.* 打算；提议
store [stɔː(r)] *v.* 存储；*n.* 商店
comprehensively [ˌkɒmprɪˈhensɪvli] *adv.* 全面地；彻底地
intermediate [ˌɪntəˈmiːdiət] *n.* 中学生；中间分子；*adj.* 中间的；*v.* 调解
instead of 代替；而不是
be represented... 由……代表
category [ˈkætəgəri] *n.* 类别；种类

occur [əˈkɜː(r)] *v.* 发生；存在于
perfection [pəˈfekʃ(ə)n] *n.* 完美；完善
primary [ˈpraɪməri] *adj.* 主要的；初级的；*n.* 初选的
correctness [kəˈrektnəs] *n.* 正确性
artificial construct and computer assistant construct 人工构造与计算机辅助构造
undesired [ˌʌndɪˈzaɪəd] *adj.* 不期望的；非意料内的
excessively [ɪkˈsesɪvli] *adv.* 过分地；过度地
leap [liːp] *n.* 跳的距离；跳跃；*v.* 跳跃；跳动
compile [kəmˈpaɪl] *v.* 编写；编制；编译
diagram [ˈdaɪəgræm] *n.* 图表；示意图；*v.* 用图表表示
quantitative [ˈkwɒntɪtətɪv] *adj.* 定量的；数量上的
node [nəʊd] *n.* 节点；[医] 结节；植物的节
terminal [ˈtɜːmɪn(ə)l] *n.* 航空站；终点站；*adj.* 终点的；致命的
solid [ˈsɒlɪd] *adj.* 固体的；结实的；*n.* 固体立方体
ellipsis [ɪˈlɪpsɪs] *n.* 省略
compile [kəmˈpaɪl] *v.* 收集；编辑；编制；编译
hypotheses [haɪˈpɒθɪsiːz] *n.* 假设；假说；猜测；猜想
precisely [prɪˈsaɪsli] *adv.* 精确地；恰好地；严谨地

·87·

Unit Thirteen
Chemical Hazard Communication

Under the provisions of the Hazard Communication Standard, employers are responsible for informing employees of the hazards and the identities of workplace chemicals to which they are exposed.

Chemical exposure may cause or contribute to many serious health effects such as heart ailments, the central nervous system, kidney and lung damage, sterility, cancer, bums, and rashes. Some chemicals may also be safety hazards and have the potential to cause fires, explosions and other serious accidents.

Because of the seriousness of these safety and health problems, and because many employers and employees know little or nothing about them, the Occupational Safety and Health Administration (OSHA) issued the Hazard Communication Standard. The basic goal of the standard is to ensure employers and employees know about work hazards and how to protect themselves; this should help to reduce the incidence of chemical source illness and injuries.

The Hazard Communication Standard establishes uniform requirements to make sure that the hazards of all chemicals imported into, produced, or used in U.S. workplaces are evaluated, and that this hazard information is transmitted to affected employers and exposed employees.

Basically, the hazard communication standard is different from other OSHA health rules because it covers all hazardous chemicals. The rule also incorporates a "downstream flow of information", which means that producers of chemicals have the primary responsibility for generating and disseminating information, whereas users of chemicals must obtain the information and transmit it to their own employees. In general, it works like this:

Chemical Manufacturers/ Importers	• Determine the hazards of each product.
Chemical Manufacturers/ Importers/Distributors	• Communicate the hazard information and associate protective measures downstream to customers through labels and MSDSs.
Employers	• Identify and list hazardous chemicals in their workplaces. • Obtain MSDSs and labels for each hazardous chemical, if not provided by the manufacturer, importer, or distributor. • Develop and implement a written hazard communication program, including labels, MSDSs, and employee training, on the list of chemicals, MSDSs and label information. • Communicate hazard information to their employees through labels, MSDSs, and formal training programs.

How Can Workplace Hazards Be Minimized?

The quality of the hazard communication program depends on the adequacy and accuracy of the assessment of hazards in the workplace. Chemical manufacturers and importers are required to review available scientific evidence concerning the hazards of the chemicals they produce or import, and to report the information they find to their employees and to employers who distribute or use their products. Downstream employers can rely on the evaluations performed by the chemical manufacturers or importers to establish the hazards of the chemicals they use. The chemical manufacturers, importers, and any employers who choose to evaluate hazards are responsible for the quality of the hazard determinations they perform. Each chemical must be evaluated for its potential to cause adverse health effects and its potential to pose physical hazards such as flammability [Definitions of hazards covered are included in the standard, see 1910.1200(c)] Chemicals that are listed in one of the following sources are to be considered hazardous in all cases:

• 29 CFR 1910, Subpart Z, Toxic and Hazardous Substances, Occupational Safety and Health Administration (OSHA), and

• Threshold Limit Values for Chemical Substances and Physical Agents in the Work

Environment, American Conference of Governmental Industrial Hygienists (ACGIH).

In addition, chemicals that have been evaluated and found to be a suspect or confirmed carcinogen in the following sources must be reported as such:

- National Toxicology Program (NTP), Annual Report on Carcinogens,
- International Agency for Research on Cancer (IARC), Monographs, and
- Regulated by OSHA as a carcinogen.

Why Is a Written Hazard Communication Program Necessary, and What Does It Include?

A written hazard communication program ensures that all employers receive the information they need to inform and train their employees properly and to design and put in place employee protection programs. It also provides necessary hazard information to employees, so they can participate in and support the protective measures in place at their workplaces.

Employers therefore must develop, implement and maintain at the workplace a written comprehensive hazard communication program that includes provisions for container labeling, collection and availability of material safety data sheets, and an employee training program. It also must contain a list of the hazardous chemicals, the means the employer will use to inform employees of the hazards of non-routine tasks (for example, the cleaning of reactor vessels), and the hazards associated with chemicals in unlabeled pipes. If the workplace has multiple employers onsite (for example, a construction site), the rule requires these employers to ensure that information regarding hazards and protective measures should be made available to the other employers onsite, where appropriate. In addition, all covered employers must have a written hazard communication program to get hazard information to their employees through labels on containers, MSDSs, and training.

The written program does not have to be lengthy or complicated, and some employers may be able to rely on existing hazard communication programs to comply with the above requirements.

The written program must be available to employees, their designated representatives, the Assistant Secretary of Labor for Occupational Safety and Health, and the Director of the National Institute for Occupational Safety and Health (NIOSH).

(Sample programs are available in the Compliance Directive CPL 2-2.38 D, Appendix E.) Also, see **Hazard Communication-A Compliance Kit** (OSHA 3104, a reference guide to step-by-step requirements for compliance with the OSHA standard). The kit can be obtained from the Government Printing Office.

How Must Chemicals Be Labelled?

Chemical manufacturers and importers must convey the hazard information they learn from their evaluations to downstream employers by means of labels on containers and material safety data sheets (MSDSs). Also, chemical manufacturers, importers, and distributors must be sure that containers of hazardous chemicals leaving the workplace are labeled, tagged, or marked with the identity of the chemical, appropriate hazard warnings, and the name and address of the manufacturer or other responsible parties. In the workplace, each container must be labeled, tagged, or marked with the identity of hazardous chemicals contained therein, and must show hazard warnings appropriate for employee protection. The hazard warning can be any type of message, words, pictures, or symbols that provide at least general information regarding the hazards of the chemical(s) in the container and the targeted organs affected, if applicable. Labels must be legible, in English (plus other languages, if desired), and prominently displayed.

Exemptions to the requirement for in-plant individual container labels are as follows:

• Employers can post signs or placards that convey the hazard information if there are a number of stationary containers within a work area that have similar contents and hazards.

• Employers can substitute various types of standard operating procedures, process sheets, batch tickets, blend tickets, and similar written materials for container labels on stationary process equipment if they contain the same information and the written materials are readily accessible to employees in the work area.

• Employers are not required to label portable containers into which hazardous chemicals are transferred from labeled containers and that are intended only for the immediate use of the employee who makes the transfer.

• Employers are not required to label pipes or piping systems.

What Are Material Safety Data Sheets, and Why Are They Needed?

The MSDS is a detailed information bulletin prepared by the manufacturer or importer of a chemical that describes the physical and chemical properties, physical and health hazards, routes of exposure, precautions for safe handling and use, emergency and first-aid procedures, and control measures.

Chemical manufacturers and importers must develop an MSDS for each hazardous chemical they produce or import, and must provide the MSDS automatically at the time of the initial shipment of a hazardous chemical to a downstream distributor or user.

Distributors also must ensure that downstream employers are similarly provided with an MSDS.

Each MSDS must be expressed in English and include information regarding the specific chemical identity of the hazardous chemical(s) involved and the common names. In addition, information must be provided on the physical and chemical characteristics of the hazardous chemical; known acute and chronic health effects and related health information; exposure limits; whether the chemical is considered to be a carcinogen by NTP, IARC or OSHA; precautionary measures; emergency and first-aid procedures; and the identification (the name, the address, and the telephone number) of the organization responsible for preparing the sheet.

Copies of the MSDS for hazardous chemicals in a given worksite are to be readily accessible to employees in that area. As a source of detailed information on hazards, they must be readily available to workers during each workshift. MSDSs have no prescribed format. ANSI standard No.Z400.1—Material Safety Data Sheet Preparation—may be used. The non-mandatory MSDS form (OSHA 174) also may be used as a guide and a copy can be obtained from OSHA field offices.

Employers must prepare a list of all hazardous chemicals in the workplace. When the list is complete, it should be checked against the collected MSDSs that the employer has been sent.

If there are hazardous chemicals used for which no MSDS has been received, the employer must contact the supplier, manufacturer, or importer to obtain the missing MSDS. A record of the contact must be maintained.

What Training is Needed to Protect Workers?

Employers must establish a training and information program for employees who are exposed to hazardous chemicals in their work area at the time of initial assignment and whenever a new hazard is introduced into their work area.

At a minimum, the discussion topics must include the following:

• The hazard communication standard and its requirements.

• The components of the hazard communication program in the employees' workplaces.

• Operations in work areas where hazardous chemicals are present.

Where the employer will keep the written hazard evaluation procedures, communications programs, lists of hazardous chemicals, and the required MSDS forms. The employees training plan must consist of the following elements:

• How the hazard communication program is implemented in that workplace, how to read and interpret the information on labels and the MSDS, and how employees can obtain and use the available hazard information.

• The hazards of the chemicals in the work area. (The hazards may be discussed by individual chemicals or by hazard categories such as flammability.)

• The measures employees can take to protect themselves from the hazards.

• Specific procedures put into effect by the employer to provide protection such as engineering controls, work practices, and the use of personal protective equipment (PPE).

• Methods and observations, such as visual appearance or smell, which workers can use to detect the presence of a hazardous chemical to which they may be exposed.

Words and Expressions

Unit Thirteen

ailment [ˈeɪlmənt] *n.* 疾病
kidney [ˈkɪdni] *n.* 肾脏
carcinogen [kɑːˈsɪnədʒən] *n.* 致癌物质
monograph [ˈmɒnəɡrɑːf] *n.* 专题著作；专题论文
legible [ˈledʒəb(ə)l] *adj.* 清晰的；易读的；易辨认的

prominently [ˈprɒmɪnəntli] *adv.* 显著地
piping system *n.* 管道系统；管路系统；管汇系统
acute [əˈkjuːt] *adj.* 急性的
chronic [ˈkrɒnɪk] *adj.* 慢性的
non-mandatory [nɒnˈmændətəri] *adj.* 非强制的

Unit Fourteen
Ergonomics Engineering

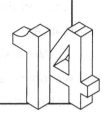

1. Introduction of Ergonomics Engineering

Ergonomics engineering (or human factors engineering) is the science of designing the job, equipment, and workplace to fit the worker. Proper ergonomic design is necessary to prevent repetitive strain injuries, which can develop over time and can lead to long-term disability. The International Ergonomics Association defines ergonomics as follows. Ergonomics engineering is the scientific discipline concerned with the understanding of interactions among humans and other elements of a system, and the profession that applies theory, principles, data and methods to design in order to optimize the human well-being and the overall system performance. Ergonomics engineering is employed to fulfill the two goals of health and productivity. It is relevant in the design of such things as safe furniture and easy-to-use interfaces to machines.

The foundations of the science of ergonomics appear to have been laid within the context of the culture of Ancient Greece. A good deal of evidence indicates that Hellenic civilization in the 5th century BC used ergonomic principles in the design of their tools, jobs, and workplaces. One outstanding example of this can be found in the description Hippocrates gave of how a surgeon's workplace should be designed and how the tools he uses should be arranged. It is also true that archaeological records of the early Egyptians Dynasties made tools, household equipment, among others that illustrated ergonomic principles.

The term "ergonomics" is derived from the Greek words ergon [work] and nomos [natural laws] and first entered the modem lexicon when Wojciech JastrzQbowski used the word in his 1857 article. Later, in the 19th century, Frederick Winslow Taylor pioneered the "Scientific Management" method, which proposed a way to find the optimum method for carrying out a given task. Taylor found that he could, for example, triple the amount of coal that workers were shoveling by incrementally reducing the size and weight of coal shovels until the fastest shoveling rate was reached.

World War II marked the development of new and complex machines and weaponry, and these made new demands on operators' cognition. The decision-making, attention, situational awareness and hand-eye coordination of the machine's operator became key in the success or failure of a task. It was observed that the fully functional aircraft, flown by the best-trained pilots, still crashed. In 1943, Alphonse Chapanis, a lieutenant in the U. S. Army, showed that this so-called "pilot error" could be greatly reduced when more logical and differentiable controls replaced confusing designs in airplane cockpits.

In the decades since the war, ergonomics has continued to flourish and diversify. The Space Age created new human factors issues such as weightlessness and extreme g-forces. How far could environments in space be tolerated, and what effects would they have on the mind and body? The dawn of the Information Age has resulted in the new ergonomics field of human-computer interaction (HCI). Likewise, the growing demand for and competition among consumer goods and electronics has resulted in more companies including human factors in product design.

At home, work, or play new problems and questions must be resolved constantly. People come in all different shapes and sizes, and with different capabilities and limitations in strength, speed, judgment, and skills. All of these factors need to be considered in the design function. To solve design problems, physiology and psychology must be included with an engineering approach.

The International Ergonomics Association (IEA) divides ergonomics broadly into three domains :

(1) Physical ergonomics: It is concerned with human anatomical, and some of the anthropometric, physiological and biomechanical characteristics as they relate to physical activities. Relevant topics include working postures, materials handling, repetitive movements, work related musculoskeletal disorders, workplace layout, safety and health.

(2) Cognitive ergonomics: It is concerned with mental processes, such as perception, memory, reasoning, and motor response, as they affect interactions

among humans and other elements of a system. Relevant topics include mental workload, decision-making, skilled performance, human-computer interaction, human reliability, work stress and training as these may relate to human-system design.

(3) Organizational ergonomics: It is concerned with the optimization of socio technical systems, including their organizational structures, policies, and processes. Relevant topics include communication, crew resource management, work design, design of working times, teamwork, participatoiy design, community ergonomics, cooperative work, new work programs, virtual organizations, telework, and quality management.

More than twenty *technical subgroups* within the Human Factors and Ergonomics Society indicate the range of applications for ergonomics. Human factors engineering continues to be successfully applied in the fields of aerospace, aging, health care, IT, product design, transportation, training, nuclear and virtual environments, among others. Kim Vicente, a University of Toronto Professor of Ergonomics, argues that the nuclear disaster in Chernobyl is attributable to plant designers not paying enough attention to human factors. "The operators were trained but the complexity of the reactor and the control panels nevertheless outstripped their ability to grasp what they were seeing during the prelude to the disaster."

Physical ergonomics is important in the medical field, particularly to those diagnosed with physiological ailments or disorders such as arthritis (both chronic and temporary) or carpal tunnel syndrome. Pressure that is insignificant or imperceptible to those unaffected by these disorders may be very painful, or render a device unusable, for those who are. Many ergonomically designed products are also used or recommended to treat or prevent such disorders, and to treat pressure-related chronic pain.

Human factors issues arise in simple systems and consumer products as well. Some examples include cellular telephones and other handheld devices that continue to shrink yet grow more complex (a phenomenon referred to as"creeping featurism"), millions of videocassette recorders (VCRs) blinking "12: 00" across the world because verify few people can figure out how to program them, or alarm clocks that allow sleepy users to inadvertently turn off the alarm when they mean to hit "snooze". A user-centered design (UCD) , also known as a systems approach or the usability engineering lifecycle, aims to improve the user-system.

2. Office Ergonomics Engineering

You may have heard of the term "ergonomics". This is sometimes referred to as human factors. Not everyone really understands what ergonomics is, what it does, or how it affects people. This leaflet will help to answer these questions and to explain how understanding ergonomics can improve health and safety in your workplace. It is aimed at anyone who has a duty to maintain and improve health and safety and who wants to gain insight into ergonomics. It gives some examples of ergonomics problems and simple, effective advice on what can be done to solve them.

Ergonomics is a science concerned with the "fit" between people and their work. It puts people first, taking account of their capabilities and limitations. Ergonomics aims to make sure that tasks, equipment, information and the environment suit each worker.

To assess the fit between a person and their work, ergonomists have to consider many aspects. These include: the job being done and the demands on the worker; the equipment used (its size, shape, and how appropriate it is for the task); the information used (how it is presented, accessed, and changed); the physical environment (temperature, humidity, lighting, noise, vibration); and the social environment (such as teamwork and supportive management).

Ergonomists consider all the physical aspects of a person, such as: the body size and shape; fitness and strength; posture; the senses, especially vision, hearing and touch; and the stresses and strains on muscles, joints, nerves. Ergonomists also consider the psychological aspects of a person, such as mental abilities, personality, knowledge, and experience.

By assessing these aspects of people, their jobs, equipment, and working environment and the interaction between them, ergonomists are able to design safe, effective and productive work systems.

Applying ergonomics to the workplace reduces the potential for accidents, reduces the potential for injury and ill health, and improves performance and productivity.

Ergonomics can reduce the likelihood of an accident. For example, in the design of control panels, designers should consider the location of switches and buttons—switches that could be accidentally knocked on or off might start the wrong sequence of events that could lead to an accident; expectations of signals and controls—most people interpret green to indicate a safe condition. If a green light is used to indicate a warning or a dangerous state, it may be ignored or overlooked; information overload—if a worker is given too much information they may become confused, make mistakes, or panic. In hazardous industries, incorrect decisions or mistaken actions have had catastrophic results.

Ergonomics can also reduce the potential for ill health at work, such as aches and pains of the wrists, shoulders and back. Consider the layout of controls and equipment; these should be positioned in relation to how they are used. Those used most often should be placed where they are easy to reach without the need for stooping, stretching or hunching. Failure to observe ergonomic principles *may have* serious repercussions, not only for individuals, but for the whole organization. Many well-known accidents might have been prevented if ergonomics had been considered in designing the jobs people did and the systems within which they worked.

Ergonomics is typically known for solving physical problems. For example, ensuring that work surfaces are high enough to allow adequate clearance for a worker's legs. *However*, ergonomics also deals with psychological and social aspects of the person and their work. For example, a workload that is too high or too low, unclear tasks, time pressures, inadequate training, and poor social support can all have negative effects on the person and the work they do.

Look for the likely causes and consider possible solutions. A minor alteration may be all that is necessary to make a task easier and safer to perform. For example:

(1) Provide height-adjustable chairs so individual operators can work at their preferred work height; Remove obstacles from under desks to create sufficient leg room;

(2) Arrange items stored on shelving so those used most frequently and those that are the heaviest are between the waist and shoulder height;

(3) Raise platforms to help operators reach badly located controls;

(4) Change shift work patterns; and introduce job rotation between different tasks to reduce physical and mental fatigue.

Unit Fourteen

Talk to employees and get them to suggest ideas and discuss possible solutions. Involve employees from the start of the process—this will help all parties to accept any proposed changes.

Always make sure that any alterations are properly evaluated by the people who do the job. Be careful that a change introduced to solve one problem does not create difficulties elsewhere.

You don't always need to consult ergonomics professionals, and the expense of making changes can often be kept low. However, you may need to ask a qualified ergonomist if you are unable to find a straightforward solution or if a problem is complex.

Health and Safety Executive (HSE) has published a range of guidance material, some of which is free. Aimed at employers and employees, this guidance provides help on how to achieve safe and healthy work environments. It includes practical evaluation checklists and advice.

A good ergonomics sense makes a good economic sense. The ergonomics input does not necessarily involve high costs, and can save money in the long term by reducing injuries and the absence from work.

An understanding of ergonomics in your workplace can improve your daily work routine. It is possible to eliminate aches, pains, and stresses at work and improve job satisfaction. Ergonomic solutions can be simple and straightforward to make even small changes such as altering the height of a chair can make a considerable difference.

 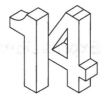

Words and Expressions — Unit Fourteen

ergonomics [ˌɜːɡəˈnɒmɪks] *n.* 工效学；人类工程学
repetitive [rɪˈpetətɪv] *adj.* 重复的；乏味的
strain [streɪn] *v.* 拉紧；损伤；*n.* 压力；劳损；扭伤
long-term *adj.* 长期的；长远的
interaction [ˌɪntərˈækʃ(ə)n] *n.* 相互影响（作用、制约、配合）；干扰（涉）
principles [ˈprɪnsəp(ə)lz] *n.* 道德原则；行为准则
well-being [ˈwel biːɪŋ] *n.* 幸福；健康
easy-to-use 易用；易于使用
context [ˈkɒntekst] *n.* 上下文；语境
illustrated [ˈɪləstreɪtɪd] *v.* 加插图于；给（书等）做图表
derived [dɪˈraɪvd] *v.* 获得；取得
lexicon [ˈleksɪkən] *v.* （某语言或学科、某人或群体使用的）全部词汇；（某学科或语言的）词汇表
pioneered [ˌpaɪəˈnɪəd] *v.* 当开拓者；做先锋；倡导
optimum method 优化方法；最佳办法
shovels [ˈʃʌv(ə)l] *v.* 铲；铲起；*n.* 铲；铁铲
weaponry [ˈwepənri] *n.* 武器；兵器
situational awareness 态势感知；情境意识
coordination [kəʊˌɔːdɪˈneɪʃn] *n.* 协作；协调
observed [əbˈzɜːvd] *v.* 看到；注意到
differentiable [ˌdɪfəˈrenʃɪəb(ə)l] *adj.* 可辨的；可区分的

airplane cockpit 飞机座舱
flourish and diversify 繁荣昌盛
weightlessness [ˈweɪtləsnəs] *n.* 失重；无重状态
domain [dəˈmeɪn] *n.* （知识、活动的）领域、范畴；（尤指旧时个人、国家等所拥有或统治的）领土、领地
anatomical [ˌænəˈtɒmɪk(ə)l] *adj.* 解剖学的；（人或动物）身体结构上的
cognitive [ˈkɒɡnətɪv] *adj.* 认知的；感知的
affect interactions 影响互动
socio-technical system 社会技术系统
participatory design 参与式设计
technical subgroups 技术小组
arthritis [ɑːˈθraɪtɪs] *n.* 关节炎
cellular [ˈseljələ(r)] *adj.* 细胞的；由细胞组成的
shrink [ʃrɪŋk] *n.* 精神病学家；心理学家；*v.* 收缩；退缩；畏缩
creeping featurism 爬行性特征
leaflet [ˈliːflət] *n.* 散页印刷品；传单；（宣传或广告）小册子；*v.* （向……）散发传单（或小册子）
assess [əˈses] *v.* 评估；评定（性质、质量）；估算；估定（数量、价值）
appropriate [əˈprəʊpriət] *adj.* 适当的；合适的；恰当的；*v.* 盗用；挪用
joints [dʒɔɪnts] *n.* 关节；（尤指构成角的）接头；接合处

Words and Expressions — Unit Fourteen

nerves [nɜːvz] *n.* 神经；神经质；*v.* 鼓足勇气；振作精神

catastrophic [ˌkætəˈstrɒfɪk] *adj.* 灾难性的；引起灾害的

layout [ˈleɪaʊt] *n.* 布局；布置

stooping, stretching or hunching 弯腰、伸展或驼背

inadequate training 训练不足

peraon *n.* 蟒蛇

waist [weɪst] *n.* 腰；腰部

straightforward [ˌstreɪtˈfɔːwəd] *adj.* 直截了当的；简单的

sense [sens] *n.* 感觉；意识；理解力；判断力；*v.* 感觉到；意识到

eliminate [ɪˈlɪmɪneɪt] *v.* 排除；清除；消灭、干掉（尤指敌人或对手）

Unit Fifteen
Health and Safety Regulation

1. Why This Guide Matters?

The Health and Safety Commission (HSC) conducted a review of health and safety regulations in 1994. It found that people were confused about the differences between guidance, Approved Codes of Practice (ACOPs) and regulations and how they relate to each other.

This document aims to explain how each fits in. It is for employers and self-employed people, but will be of interest to anyone who wants to know how health and safety law is meant to work.

2. What the Health and Safety Law Requires?

The basis of the British health and safety law is the Health and Safety at Work Act 1974.

The Act sets out the general duties which employers have towards employees and members of the public, and employees have to themselves and to each other.

These duties are qualified in the Act by the principle of "so far as is reasonably practicable". In other words, an employer does not have to take measures to avoid or reduce the risk if they are technically impossible or if the time, trouble or cost of the measures would be grossly disproportionate to the risk.

What the law requires here is what good management and common sense would lead employers to do anyway: that is, to look at what the risks are and take sensible measures to tackle them.

The Management of Health and Safety at Work Regulations 1999 (the Management Regulations) generally makes more explicit what employers are required to do to manage health and safety under the Health and Safety at Work Act. Like the Act, they apply to every work activity.

The main requirement on employers is to carry out a risk assessment. Employers with five or more employees need to record the significant findings of the risk assessment.

The risk assessment should be straightforward in a simple workplace such as a typical office. It should only be complicated if it deals with serious hazards such as those on a nuclear power station, a chemical plant, a laboratory or an oil rig.

The HSE leaflet Five Steps to Risk Assessment will give you more information. Besides carrying out a risk assessment, employers also need to:

—make arrangements for implementing the health and safety measures identified as necessary by the risk assessment;

—appoint competent people (often themselves or company colleagues) to help them to implement the arrangements;

—set up emergency procedures;

—provide clear information and training to employees;

—work together with other employers sharing the same workplace.

Other regulations require action in response to particular hazards, or in industries where hazards are particularly high.

3. European Law

In recent years much of Britain's health and safety law has originated in Europe. Proposals from the European Commission may be agreed by Member States, who are then responsible for making them part of their domestic law.

Modern health and safety law in this country, including much of that from Europe, is based on the principle of risk assessment described above.

4. Action on Health and Safety

The Health and Safety Commission and its operating arm, the Executive (HSC/E), have spent over twenty years modernizing the structure of health and safety law. Their aims are to protect the health, safety and welfare of employees, and to safeguard others, principally the public, who may be exposed to risks from work activity.

HSC/E consult fully with people affected by their legislative proposals, and adopt

→ Health and Safety Regulation

various approaches based on assessing and controlling risk.

Among the things that can prompt action from HSC/E are;

—changes in technologies, industries or risks;

—evidence of accidents and ill health, plus public concern;

—European Directives.

Where HSC/E consider action is necessary to supplement existing arrangements, their three main options are:

—guidance;

—Approved Codes of Practice; and

—regulations.

HSC/E try to take whichever option, or options to allow employers most flexibility and cost them least, while providing proper safeguards for employees and the public.

5. Guidance

HSC/E published guidance on a range of subjects (please see the end of this guide).

Guidance can be specific to the health and safety problems of an industry or of a particular process used in a number of industries.

The main purposes of the guidance are:

—to interpret—helping people to understand what the law says—including for example how requirements based on EC Directives fit with those under the Health and Safety at Work Act;

—to help people comply with the law;

—to give technical advice.

Following guidance is not compulsory and employers are free to take other action. But if they do follow the guidance they will normally be doing enough to comply with the law. (Please also see the sections below on Approved Codes of Practice and regulations, which explain other ways in which employers are helped to know whether they are doing what the law requires.)

HSC/E aim to keep the guidance up-to-date, because as technologies change, risks and the measures needed to address them change too.

15 →→ Unit Fifteen

6. Approved Codes of Practice

Approved Codes of Practice offers practical examples of good practice.

They give advice on how to comply with the law by, for example, providing a guide to what is "reasonably practicable" . For example, if regulations use words like " suitable and sufficient", Approved Code of Practice can illustrate what this requires in particular circumstances.

Approved Codes of Practice has a special legal status. If employers are prosecuted for a breach of health and safety law, and it is proved that they have not followed the relevant provisions of the Approved Code of Practice, a court can find them at fault unless they can show that they have complied with the law in some other way.

HSC consulted in 1995 on the role of Approved Codes of Practice in the health and safety system and concluded that they could still be used in support of legal duties in specific circumstances.

7. Regulations

Regulations are law, approved by Parliament. These are usually made under the Health and Safety at Work Act, following proposals from HSC. This applies to regulations based on EC Directives as well as " home-grown" ones.

The Health and Safety at Work Act, and general duties in the Management Regulations, are goal-setting (see "What form do they take?") and leave employers freedom to decide how to control risks which they identify. Guidance and Approved Codes of Practice give advice. But some risks are so great, or the proper control measures so costly, that it would not be appropriate to leave employers discretion in deciding what to do about them. Regulations identify these risks and set out specific action that must be taken. Often these requirements are absolute—to do something without qualification by whether it is reasonably practicable.

8. Application

Some regulations apply across all companies, such as the Manual Handling Regulations, which apply wherever things are moved by hand or bodily force, and the Display Screen Equipment Regulations, which apply wherever VDUs are used. Other regulations apply to hazards unique to specific industries, such as mining or nuclear.

9. Form

HSC will where appropriate propose regulations in the goal-setting form, that is, setting out what must be achieved, but not how it must be done.

Sometimes it is necessary to be prescriptive, that is spelling out in detail what should be done. Some standards are absolute. For example, all mines should have two exits; contacts with live electrical conductors should be avoided. Sometimes European law requires prescription.

Some activities or substances are so inherently hazardous that they require licensing, for example, explosives and asbestos removal. Certain big and complex installations or operations require "safety cases", which are large-scale risk assessments subject to scrutiny by the regulator. For example, railway companies are required to produce safety cases for their operations.

10. The Relationship Between the Regulator and the Industry

As mentioned above, HSC consults widely with those affected by its proposals.

HSC/E work through:

—HSC's Industry and Subject Advisory Committees, which have members drawn from the areas of work they cover, and focus on health and safety issues in particular industries (such as the textile industry, construction and education or areas such as toxic substances and genetic modification);

—intermediaries, such as small firms organisations;

—providing information and advice to employers and others with responsibilities under the Health and Safety at Work Act;

—guidance to enforcers, both HSE inspectors and those of local authorities;

—the day-to-day contact which inspectors have with people at work.

HSC directly canvasses the views of small businesses. It also seeks views in detail from representatives of small businesses about the impact on them of proposed legislation.

Words and Expressions — Unit Fifteen

Health and Safety Commission 健康安全委员会
health and safety regulation 健康与安全章程
approved codes of practice 实施规范
guidance [ˈɡaɪdns] n. 指南；指导
regulation [ˌreɡjuˈleɪʃn] 法规；条例
relate to 涉及
fit in （相）适应；适合
employer [ɪmˈplɔɪə(r)] n. 雇主；老板
employee [ɪmˈplɔɪiː] n. 雇员；从业员工
self-employed people 自由职业者；个体职业者
of interest to 使……，对……
set out 陈述
common sense 常识
apply to 适用于；应用于
carry out 执行；实行
risk assessment 风险评估
deal with 处理；涉及
nuclear power station 核电站
chemical plant 化工厂
oil rig 石油钻机；钻油平台
make arrangements for 安排某事
competent person 称职人员；有资格人士
emergency procedure 紧急程序
in response to 响应；回答
originate in 起源于（发生于）；源自
European Commission 欧洲委员会
member state 成员国
domestic law 国内法

executive [ɪɡˈzekjətɪv] n. 执行者
the public 公众；民众
be exposed to 曝光；暴露
legislative proposal 立法建议
controlling risk 控制风险
approach [əˈprəʊtʃ] 方法；途径
a range of 一系列；一些；一套
specific to 针对；特定于
fit with 符合
comply with 遵守
technical advice 技术咨询；技术建议
good practice 好的做法；好的实践
reasonably practicable 合理可行
legal status 法律地位
relevant provision 有关规定；有关条文
be used in 用于
legal duty 法律职责
parliament [ˈpɑːləmənt] n. 议会；国会
as well as 也；和……一样；不但……而且
management regulation 管理规定
be appropriate to 适当；合适
VDUs n. 显示装置；显示器
live electrical conductor 带电导体
inherently hazard 固有的危险
subject to 使服从；使遭受；受……管制
as mentioned above 如上所述
consult [kənˈsʌlt] n. 顾问
advisory committee 咨询委员会
impact on 影响；对……冲击

Unit Sixteen
Emergency Management

1. Introduction to Disasters and Emergency Management

A disaster is a state in which a population, a population group, or an individual is unable to cope with or overcome the adverse effects of an extreme event without outside help. The impact of an extreme event may include significant physical damage or destruction, loss of life, or drastic changes to the environment. It is a phenomenon that can cause damage to life and property and destroy the economic, social, and cultural life of people.

The above definition perceives disasters as the consequence of an inappropriately managed risk. For example, when looking at an extreme event such as a flood, although the primary cause for a flood is extreme rainfall, snowmelt, or a combination of both, the impact or magnitude of a flood is determined by human influences. The risks associated with a disaster are a function of both the hazards and the vulnerability of the affected group. Hazards that strike in areas with low vulnerability will never become disasters, as is the case with uninhabited regions. Developing countries suffer the greatest costs when a disaster hits; more than 95% of all deaths caused by disasters occur in developing countries, and losses due to natural disasters are 20 times greater as a percentage of Gross Domestic Product (GDP) in developing countries than in industrialized countries.

Disaster management is defined as the organization and management of resources and responsibilities for dealing with all humanitarian aspects of emergencies, in particular preparedness, mitigation, response, and recovery in order to lessen the impact of disasters. It deals specifically with the processes used to protect populations and/or organizations from the consequences of disasters. However, it does not necessarily extend to the prevention or elimination of the threats themselves, although the study and prediction of threats is an imminent part of disaster management. International organizations focus on community-based disaster preparedness, which assists communities to reduce their vulnerability to disasters and strengthen their coping capacities.

When the capacity of a community or country to respond to and recover from a disaster is exceeded, outside help is necessary. This assistance comes from different sources including government and nongovernment organizations. The definition of disaster as a state where people at risk can no longer help themselves conforms to the modern view of a disaster as a social event, where the people at risk are vulnerable to the effects of an extreme event because of their social conditions.

According to this view, disaster management is not only a technical task, but also a social task. Therefore, community-based disaster preparedness becomes an integral part of disaster management. By reducing a community's vulnerability to a disaster and by building upon its coping capacities, skills, and resources, these communities are better able to meet future crises.

The first people to respond to a disaster are those living in the local community, and they are also the first to start rescue and relief operations. They know what their needs are, have an intimate knowledge of the area, and may have experienced similar events in the past. Therefore, community members need to be consulted and involved in the response and recovery operations, including assessment, planning, and implementation. Consultation can take place through community leaders, representatives of women's or other community associations, beneficiaries, and other groups.

2. Types of Disasters

Disasters can be classified into two subcategories: natural hazards and technological or man-made hazards. Natural hazards are naturally occurring physical phenomena caused by either rapid or slow onset events that can be geophysical, hydrological, climatological, meteorological, and/or biological in nature. Geophysical disasters include earthquakes, landslides, tsunamis, and volcanic activity; hydrological disasters include hazards such as avalanches and floods; climatological disasters include hazards such as extreme temperatures, droughts, and wildfires; meteorological disasters include cyclones, hurricanes, storms, and/or wave surges;

and biological disasters include disease epidemics and infestations such as insect and/or animal plagues. Some natural disasters can result from a combination of different hazards, for example, floods can be the result of tsunamis, storm surges, hurricanes, or cyclones or a combination of all four. However, after a flood, epidemics such as cholera, malaria, and dengue fever begin to emerge.

Technological or man-made hazards are events that are caused by humans and occur in or close proximity to human settlements. This can include environmental degradation, pollution, complex emergencies and/or conflicts, cyber-attacks, famine, displaced populations, industrial accidents, and transport accidents. Workplace fires are more common and can cause significant property damage and loss of life. Communities are also vulnerable to threats posed by extremist groups who use violence against both people and property. High-risk targets include military and civilian government facilities, international airports, large cities, and high-profile landmarks. Cyberterrorism involves attacks against computers and networks done to intimidate or coerce a government or its people for political or social objectives.

Technological disasters are complex by their very nature and could include a combination of both natural and man-made hazards. In addition, there is a range of challenges such as climate change, unplanned urbanization, underdevelopment and poverty, as well as the threat of pandemics that will shape disaster management in the future. These aggravating factors will result in the increased frequency, complexity, and severity of disasters.

Climate change ranks among the greatest global problems of the 21st century, and the scientific evidence on climate change is stronger than ever. For example, the Intergovernmental Panel on Climate Change (IPCC) released its Fourth Assessment Report in early 2007, saying that climate change is now unequivocal. It confirms that extremes are on the rise and that the most vulnerable people, particularly in developing countries, face the brunt of impacts. The gradual expected temperature rise may seem limited, with a likely range from 20°C to 40°C predicted for the coming century. However, a slightly higher temperature is only an indicator that is much more skewed. Along with the rising temperature, we can experience an increase in both the frequency and intensity of extreme weather events such as prolonged droughts, floods, landslides, heat waves, and more intense storms; the spreading of insect-borne diseases such as malaria and dengue to new places where people are less immune to them; a decrease in crop yields in some areas due to extreme droughts or downpours and changes in timing and reliability of rainfall seasons; a global sea level rise of several centimeters per decade, which will affect coastal flooding, water supplies, tourism, and fisheries, and tens of millions of people will be forced to move inland; and melting glaciers, leading to water supply shortages. Climate change is here to stay and will accelerate. Although climate change is a global issue with impacts all over the world, those people with the least resources have the least capacity to adapt and therefore are the most vulnerable. Developing countries, more specifically its poorest inhabitants, do not have the means to cope with floods and other natural disasters; to make matters worse, their economies tend to be based on climate and/or weather-sensitive sectors such as agriculture and fisheries, which makes them all the more vulnerable.

The impact of underdevelopment, unplanned urbanization, and climate change is present in our everyday work; disasters are a development and humanitarian concern. A considerable incentive for rethinking disaster risks as an integral part of the development process comes from the aim of achieving the goals laid out in the Millennium Declaration. The Declaration sets forth a road map for human development supported by 191 nations. Eight Millennium Development Goals (MDGs) were agreed upon in 2000, which in turn have been broken down into 18 targets with 48 indicators for progress. Most goals are set for achievement by 2015. The MDGs have provided a focus for development efforts globally. While poverty has fallen and social indicators have improved, most countries will not meet the MDGs by 2015, and the existing gap between the rich and the poor will widen. Recently, the campaigns on poverty have resulted in key milestones on aid and debt relief. While positive, much more is needed if the MDGs are to become a reality. These efforts to reduce poverty are vital for vulnerability reduction and strengthened resilience of communities to disasters.

Today 50% of the world's population live in urban centers, and by 2030 this is expected to increase to 60%. The majority of the largest cities, known as megacities, are in developing countries, where 90% of the population growth is urban in nature. Migration from rural to urban areas is often triggered by repeated natural disasters and the lack of livelihood opportunities. However, at the same time many megacities are built in areas where there is a heightened risk for earthquakes, floods, landslides, and other natural disasters. Many people living in large urban centers such as "slums" lack access to improved water, sanitation, security of tenure, durability of housing, and a sufficient living area. This lack of access to basic services and livelihood leads to an increasing risk of discrimination, social exclusion, and ultimately violence.

3. Emergency Planning

The terrorist attacks of September 11, 2001, illustrated the need for all levels of government, the private sector, and nongovernimental agencies to prepare for, protect against, respond to, and recover from a wide spectrum of events that exceed the capabilities of any single entity. These events require a unified and coordinated approach to planning and emergency management.

Knowing how to plan for disasters is critical in emergency management. Planning can make the difference in mitigating against the effects of a disaster, including saving lives and protecting property and helping a community recover more quickly from a disaster. Developing an effective Emergency Operations Plan (EOP) can have certain benefits for a municipality including the successful evacuation of its citizens and the ability to survive on its own without outside assistance for several days. The consequences of not having an emergency plan include the following: The need for immediate assistance and a higher number of casualties resulting from an attempted evacuation, with them being ineligible for the full amount of aid from upper-tier municipalities (regions or counties) , provincial, state, or federal governments. In the United States, counties that do not have emergency plans cannot declare an emergency and are ineligible for any aid or for the full amount of aid. In Ontario, Canada, all municipalities must have an emergency plan in accordance with the Emergency Management and Civil Protection Act.

Emergency planning is not a one-time event. Rather, and it is a continual cycle of planning, training, exercising, and revision that takes place throughout the five phases of the emergency management cycle: preparedness, prevention, response, recovery, and mitigation. The planning process does have one purpose: the development and maintenance of an up-to-date EOP. An EOP can be defined as a document describing how citizens and property will be protected in a disaster or emergency. Although the emergency planning process is cyclic, EOP development has a definite starting point. There are four steps in the emergency planning process: the hazard analysis, the development of an EOP, testing the plan through a series of training exercises, and plan maintenance and revision.

Step 1: The hazard analysis. It is the process by which hazards that threaten the community are identified, researched, and ranked according to the risks they pose and the areas and infrastructure that are vulnerable to damage from an event involving the hazards. The outcome of this step is a written hazard analysis that quantifies the overall risk to the community from each hazard. The hazard analysis is a component of the emergency planning process.

Step 2: The development of an EOP. The EOP includes the basic plan, functional annexes, hazard-specific appendices, and implementing instructions. The outcome of this step is a completed plan, which is ready to be trained, exercised, and revised based on lessons learned from the exercises.

Step 3: Testing the plan through a series of training exercises. Training exercises of different types and varying complexity allow emergency managers to see what in the plan is unclear and what does not work. The outcomes of this step are lessons learned about weaknesses in the plan that can then be addressed in Step 4.

Step 4: Plan maintenance and revision. The outcome of this step is a revised EOP, based on current needs and resources, which may have changed since the development of the original EOP. After the EOP is developed, Step 3 and Step 4 repeat in a continual cycle to keep the plan up to date. If you become aware that your community faces a new threat such as terrorism, the planning team will need to revisit Step 1 and Step 2. Emergency planning is a learning effort because the disaster response requires coordination between many community agencies and organizations and different levels of government. Furthermore, different types of emergencies require different kinds of expertise and response capabilities. Thus, the first step in emergency planning is the identification of all of the parties that should be involved.

Obviously, the specific individuals and organizations involved in response to an emergency will depend on the type of disaster. Law enforcement will probably have a role to play in most events, as will fire, Emergency Medical Services (EMS), voluntary agencies (i.e., the Red Cross) , and the media. On the other hand, the hazardous materials personnel may or may not be involved in a given incident but should be involved in the planning process because they have specialized expertise that may be called on.

Getting all stakeholders to take an active interest in emergency planning can be a daunting task. To schedule meetings with so many participants may be even more difficult. It is critical, however, to have everyone's participation in the planning process and to have them feel ownership in the plan by involving them from the beginning. Also, their expertise and knowledge of their organizations' resources are crucial to developing an accurate plan that considers the entire community's needs and the resources that could be made available in an emergency.

It is definitely in the community's best interest to have the active participation of all stakeholders. The following are recommendations of what can be done to ensure that all stakeholders that should participate in the discussions do, so that a plan is formulated. Give the planning: Learn plenty of notice of where and when the planning meeting will be held. If time permits, you might even survey the team members to find the time and place that will work for them. Provide information about team expectations ahead of time. Explain why participating on the planning team is important to the participants' agency and to the community itself. Show the participants how they will contribute to a more effective emergency response. Ask the Chief Administrative Officer (CAO) or their Chief of Staff to sign the meeting announcement. A directive from the executive office will carry the authority of the CAO and send a clear signal that the participants are expected to attend and that emergency planning is important to the community. Allow for flexibility in scheduling after the first meeting. Not all team members will need to attend all meetings. Some of the work can be completed by task forces or subcommittees. Where this is the case, gain concurrence on time frames and milestones but let the subcommittee members determine when it is most convenient to meet. In addition, emergency managers may wish to speak with their colleagues from adjacent municipalities to gain their ideas and inputs on how to gain and maintain interest in the planning process.

16 → → Unit Sixteen

Working with the personnel from other agencies and organizations requires collaboration. Collaboration is the process by which people work together as a team toward a common goal, in this case, the development of a community EOP. Successful collaboration requires a commitment to participate in shared decision making; a willingness to share information, resources, and tasks; and a professional sense of respect for individual team members. Collaboration can be made difficult by differences among agencies and organizations in terminology, experience, mission, and culture. It requires the flexibility from team members to reach an agreement on common terms and priorities and humility to learn from others' ways of doing things. Also, collaboration among the planning team members benefits the community by strengthening the overall response to the disaster. For example, collaboration can eliminate duplication of services, resulting in a more efficient response and expanding resource availability. It can further enhance problem solving through cross-pollination of ideas.

adverse [ˈædvɜːs] *adj.* 不利的；有害的；反面的

destruction [dɪˈstrʌkʃ(ə)n] *n.* 摧毁；破坏

vulnerability [ˌvʌlnərəˈbɪləti] *n.* 脆弱性；弱点

mitigation [ˌmɪtɪˈgeɪʃn] *n.* 减轻；缓和

assessment [əˈsesmənt] *n.* 评定；估价

implementation [ˌɪmplɪmenˈteɪʃ(ə)n] *n.* 实现；履行

maintenance [ˈmeɪntənəns] *n.* 维护；维持

epidemic [ˌepɪˈdemɪk] *n.* 流行病；蔓延

proximity to 接近；临近

Civil Protection Act 民事保护法案

emergency planning 应急预案

infrastructure [ˈɪnfrəstrʌktʃə(r)] *n.* 基础设施；公共建设

Unit Seventeen

Construction Safety

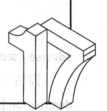

Construction safety is a global issue in that it is a concern wherever construction activities take place. The reality is that the construction industry continually has injury and fatality statistics that make it one of the most dangerous industries in which to work. Even though tremendous improvements have been made in safety performance in some countries, the construction industry continues to lag behind most other industries. This has been the experience within most countries. As the world has become smaller through technology and through cooperative arrangements that cross many borders, the issue of construction worker safety has become a well-recognized problem and represents a concern that is shared worldwide.

The construction safety problems that exist are rarely unique to a single country. In the global market, construction problems are very similar from country to country. This is quite evident when attending international construction safety conferences where the themes of primary interest have general appeal to all participants. Since construction safety problems appear to be ubiquitous, this also means that the problems of construction safety can be addressed and solved on a global scale, resulting in improvements that can be observed on a global scale. Thus, solutions to safety problems in one country can readily be adopted in other countries to generate further improvements.

Professional engineers are dedicated to protecting the public safety, and many of our members are in positions of "trust" in our nation's building departments where they struggle to provide codes that minimize the public's exposure to accidental injuries. The problem is twofold in that the written applicable regulations, codes, standards, and safety rules, such as those provided by government OSFIA, industry crane manufacturers, and professional societies provide the structure while enforcement and training must support project installation procedures. Construction engineering practitioners are facing a dilemma in safeguarding workers and the community affected by a building project. These engineers, who represent owners through their managers and contractors, have installed a series of safe standards that generally lead to a low injury rate on a project; owners must participate actively in project safety including the assignment of the contractor's responsibility to prepare a site safety plan that includes provisions for crane and hoisting safety. The owner must require compliance with the applicable regulations, codes, and standards.

"The devil is in the details," and the engineering community is being held liable for many injuries that are caused by human error. Recent examples include a crane accident possibly caused by faulty nylon straps that failed due to either overuse, misuse, or abuse to temporarily support the six-ton bracing collar during the jump of a 200-foot-tall tower crane. OSHA has announced that they will release a proposed revision of its crane and derrick standard. Having a crane manufacturer representative, a master rigger, and a safety specialist on site during a jump may help. Adding layers of supervisors may also be counterproductive; the responsible parties must educate their employees and supply them with safe tools and supports. The architect/engineer, as an action agent for the owners' contract documents, has a responsibility to include contract requirements that specify regulations, codes, standards, and sequences such as ASCE's policy statements on construction site safety, crane safety, and their crane safety manual, while the remainder of the team is responsible for implementing the proper steps in accordance with the required specifications. Another example is the case where a crane's swing table separated from its ring support. Earlier, a cracked part had been repaired by welding and reinstalled; the catastrophic failure of the crane occurred while operating. The public must be protected during crane and hoisting operations.

The suggestion that employers observe workers and provide feedback to them on both safe practices and at-risk actions has been used on some sites and is known as behavioral-based safety. Anonymous safety records for workers that are transferable to other sites and contractors may be a method of improving safety awareness provided that it does not add to the paper backlog. It seems like each suggested solution has many negatives that are not desirable. The responsibility for worksite safety must be assigned to all the team members, which includes government and private owners, insurers, designers, and contractors. The worker that steps on an unsafe scaffold—one that he or she built—will often seek damages from the entire project participants. Perhaps, if a special safety policy group were set up on the site to monitor safety actions and damages, then projects could avoid some of the pitfalls that lead to serious accidents. Cameras are common on a construction site to record the daily progress; they should be upgraded to record much of the site work activity in detail. Each accident, no matter how small or large, should be investigated to learn how to avoid it in the future. In such cases a film of the activity in the vicinity of the accident would be of great help. The owner must assign safety as a responsibility to his primary contractor authority and this agent must coordinate site safety activities, including the management of crane and hoisting operations.

A number of serious failures have occurred in steel structures, including the Minneapolis 1–35 W bridge disaster. Connections are the most common cause of claims and failures and they are the most difficult part of steel design. Many codes assign responsibility for their suitability to the structural engineer, though it is standard practice for the structural engineer of record to designate the connection design to a professional engineer working for the steel fabricator. In most cases the structural engineers conduct peer reviews of their work in accordance with recommendations of various structural associations and as required by many insurance companies representing the structural design engineer. Building codes are being revised in an effort to increase a building's resistance to progressive collapse through stronger horizontal ties. Mandated peer reviews may correct some of these potential hazards.

Progressive failures that result from destructive fires leave unanswered many questions regarding the speed and extent of the damage in a steel-framed building. The fireworks display ignited the flammable insulation layer after penetrating the exterior metal facade at several points. It then spread to the interior of the building where debris and decorations provided fuel for accelerating the flames. In addition fireworks landing on the roof spread fire to the lower floors where high winds swept it further to debris piles that ignited very quickly. Chimney effects in the hollow core between the 5th and 25th floors probably accelerated the fires as they spread out from the central core. The heat was so severe that it partially collapsed the core during its passage. It is unknown whether the building will be torn down or if renovations can save the structure's steel skeleton that is still standing.

As engineers representing either the government, owners, designers, or contractors, we must work together to prohibit the stockpiling of debris on construction sites and consider the effects of an unexpected fire on a project. We must minimize the use of materials that are flammable and carefully detail the fireproof closure of shafts or cores during a fire emergency while the project is under construction or during its intended life. Perhaps we should consider revising our building codes and require that the sprinkler system be pressurized and operable as early as possible in a project's construction.

Failures that lead to serious injuries have a common thread that passes from crane accidents to scaffold failures to steel connections and to all kinds of construction accidents. The chain is no stronger than its weakest link, and to strengthen the weak links we must as a profession mandate the use of cameras and the resulting photographic record to instruct workers on what appears to be currently unsafe practices, to encourage OSHA to issue updated regulations, and push employers to train their associates—such as the field representatives of the contractor and of the design engineer—using the most current safety requirements.

Supervisors must audit and check the jobsite procedures during crane jumps. Riggers must follow the manufacturers' instructions and not take shortcuts: In one crane failure the jump crew used half of the eight recommended attachment points and used synthetic slings at an angle that led to an unorthodox choker, decreasing holding capacity and exposing them to sharp edges. The consequences of taking shortcuts on a construction site may lead to a disaster. Construction teams and crews do not always perform as expected; qualified rigging professionals must inspect each step. Registered professional engineers should approve the manufacturers' instructions and the overall plans for the construction site. The owner's safety agent must develop a site-specific crane and rigging safety plan, which includes production and critical lifts, and this safety authority must require the training of the management staff and the jobsite personnel in crane and rigging safety procedures.

Construction work crews must be well-trained in safe practices and engineers

Unit Seventeen

must continue to require inspection procedures that mimic jobsite conditions whenever the public is at risk. Engineer approval should be based on test lifts that are similar to jobsite conditions and utilize the site equipment. We, as engineers representing the government, owners, designers, or contractors, must do more to overcome careless actions on construction projects. The engineers' vow to protect the public safety requires recognition that having the best system of construction codes and standards does not prevent careless or inconsiderate construction participants from finding a way to introduce unsafe practices, such as contacting electrical power line hazards on a project or an adjacent site.

Experimental data indicates that employees learn more when they are required to participate in interactive problem solving, like installing a model crane jump or a scaffold. Crews that were required to build devices after having studied the standards, including the manufacturers' installation instructions, demonstrated a better grasp of the issues than the ones who learned only through lectures, reading, and tests. Contractors are encouraged to develop a site-specific safety plan and job hazard analysis for common field activities undertaken by their construction crews. Each craft has risky exposures that may be avoided through practice on safety models. OSHA has developed safety-training courses for both craft workers and managers that should be helpful on most projects. Manufacturers must standardize load chart formats and equipment control configurations with all manuals written for the end user. Documentation should emphasize equipment limitations for wind and water operation.

The major risks associated with crane and hoisting operations demand an increased effort to improve crane and rigging safety by the construction team.

Words and Expressions — Unit Seventeen

take place 举行；发生
fatality [fəˈtæləti] *n.* 死亡；宿命；灾祸
dedicate to 奉献；从事于；献身于
contractor [kənˈtræktə(r)] *n.* 订约人；承包人；收缩物
crane [kreɪn] *n.* 鹤；起重机；伸长
rigger [ˈrɪɡə(r)] *n.* 索具装配人；机身装配员；暗中操纵者
counterproductive [ˌkaʊntəprəˈdʌktɪv] *adj.* 反生产的；使达不到预期目标的

site safety 工地安全
scaffold [ˈskæfəʊld] *v.* 搭脚手架于（某处）；使站在脚手架上；把……处死刑
supervisor [ˈsuːpəvaɪzə(r)] *n.* 监督人；[管理]管理人；检查员
hazard [ˈhæzəd] *n.* 危险；冒险

Unit Eighteen

Safety Management

What is the Safety Profession?

The primary focus for the safety profession is the prevention of harm to people, property and the environment. Safety professionals apply principles drawn from such disciplines as engineering, education, psychology, physiology, enforcement, hygiene, health, physics and management. They use appropriate methods and techniques of loss prevention and loss control. "Safety science" is a twenty-fist century term for everything that goes into the prevention of accidents, illnesses, fires, explosions and other events which harm people, property and the environment.

The U. S. has a lot to gain by reducing the number of these preventable events. The National Safety Council estimated that in the U. S. accidents alone cost over $ 480. 5 billion in 1998. Fire-related losses exceed for $ 8 billion per year.

Illnesses caused by exposing people to harmful biological, physical and chemical agents produce great losses each year and accurate estimates of their impact are hard to make.

In addition, pollution of all kinds causes damage to all forms of life. This generates skyrocketing cleanup costs and threatens the future habitability of our planet.

The term "safety science" may sound new, but many sources of safety science knowledge are hundreds of years old. All of the following are knowledge areas of safety science:

Chemistry and biology provide knowledge about hazardous substances.

Physics tells people about electricity, heat, radiation and other forms of energy that must be controlled to ensure safe use.

Ergonomics helps people understand the performance limits of humans and helps them design tasks, machines, work stations and facilities which improve performance and safety.

Psychology helps people understand human behaviors that can lead to or avoid accidents.

Physiology, biomechanics and medicine help people understand the mechanisms of injuries and illnesses and how to prevent them.

Engineering, business management, economics, and geology give people the knowledge necessary to improve safety in our society.

The things that can cause or contribute to accidents, illnesses, fire and explosions, and similar undesired events are called "hazards". Safety science gives people the ability to identify, evaluate, and control or prevent these hazards. Safety science provides management methods for setting policies and securing funds to operate safety activities in a company.

Hazard control activities go on every day throughout the world. From the careful design and operation of nuclear power generating stations to the elimination of lead-based paints in homes, the efforts to reduce threats to public safety go on nonstop. The application of safety science principles occurs in many places: in the workplace, in all modes of transportation, in laboratories, schools, and hospitals, at construction sites, on oil drilling rigs at sea, in underground mines, in the busiest cities, in the space program, on farms, and anywhere else where people may be exposed to hazards.

Safety science helps people understand how something can act as a hazard. People must know how and when the hazard can produce harm and the best ways to eliminate or reduce the danger. If a hazard cannot be eliminated, we must know how to minimize exposures to the hazard. This costs money and requires assistance from designers, owners and managers. Safety professionals must know the most cost-effective ways to reduce the risk and how to advise employees, owners and managers. By applying safety science, all of these activities can be effectively carried out. Without safety science, safety professionals rely on guesswork, mythology and superstition.

Safety professionals are the specialists in the fight to control hazards. To be called professions, they must acquire the essential knowledge of safety science through education and experience so that others can rely on their judgments and recommendations.

Top safety professionals demonstrate their competence through professional certification examinations. Regardless of the industry, safety professionals help achieve safety in the workplace by identifying and analyzing hazards which potentially create injury and illness problems, developing and applying hazard controls, communicating safety and health information, measuring the effectiveness of controls, and performing follow-up evaluations to measure the continuing improvement in programs.

Safety Management System

Depending on the perspective taken, there are multiple definitions of a safety management system, but its definition is always concerned with three core issues, "safety" "management" and "system". Safety refers to its opposite: accident losses or risks. Management connects accident causes to organizational control and actions. The system refers to a systematic frame work or models that provide the logic of safety management. To sum up, an SMS (safety management system) means a system containing management principles and activities for controlling risks and preventing accidents.

Depending on their background, SMSs are either narrowly or broadly defined and developed, each with its own pros and cons. Some provide a definition that is directly based on their own industrial activity or even operational SMSs; their angle is practical and meant to achieve the desired safety performance or meet specific safety policies. Others are more abstract in their definitions of an SMS whereby its constituent parts are elaborated along the lines of traditional management systems directed at the continuous improvement of safety performance. Despite the fact that the content of SMSs always pertains to activities, processes, documented procedures or functional control systems, a clear delineation of an SMS is imperative for its implementation as it determines the required resources as well as the responsibilities of the SMS. An SMS is essentially a mechanism that can be designed in different ways apart from its environment, such as (safety) culture or a certain industrial context. In our overview, the definition of an SMS makes it possible to distinguish it from other such management systems.

Safety management developed along with the improvement of safety theories, practices and standards. An SMS is primarily driven by accident analysis and prevention. Even laws, regulations and standards are prompted by accidents because their consequences raise the public's awareness of safety and their acceptance of risk: as low as possible in a practical sense. The history of safety management also shows increased attention for economic reasons with respect to the development of SMS. Indeed, an effective SMS plays an important role in the assessment of a company's creditworthiness and its ability to control risk (e. g. through insurances). The overview of the history of SMS' development has shown that safety management systems can significantly contribute to the improvement of organizational management as a whole.

The theoretical modelling of SMSs can improve the effectiveness and efficiency of SMS' development. Overall there are three main groups of models: (1) Accident theories and models describe the events and cause-effect relationships. They provide the means to develop scenarios for risk analysis; (2) Safety barriers inserted in the event sequences are the connection between the accident model and the management model. The barriers show the elaborate ways that safety management systems have for controlling accidents; (3) The management models are important as they show how the safety barriers are to be managed. Subsequently, the risk is controlled. The hierarchical models only show the framework of management, but it is difficult to make sure that the safety systems and barriers are functioning as designed. Therefore, factors that influence the risk or barrier failure receive increasing research effort. In terms of a complete SMS, the events model, the events model with barriers inserted and the management model are the three stages for modelling and still three important topics for safety management research.

In accordance with the purpose of setting up an SMS and carrying out research into it, control and compliance are critical. Either at a theoretical level or at a practical level, SMSs are designed to control unwanted events with a high probability or loss. The PDCA control loop is a central idea applied in safety management systems and all its sub-systems. Controls, techniques and data analysis are the main concerns in these sub-systems. In practice, SMSs are popular for their role in compliance management. This explains why obtaining a safety certificate can sometimes motivate companies to continually improve their SMSs. According to the literature, an

18 →→ Unit Eighteen

integrated management system is more advanced than independent safety systems, as safety is just one of the comprehensive organization management objectives. In terms of purpose, control is the obvious aim of an SMS for which some functions to prevent accidents need to be fulfilled; a standard complied SMS is the necessary requirement in a global market. The demand for safety of companies ultimately determines the purpose of their SMSs.

Elements of SMSs have a bearing on the definition of safety management, modelling and the actual purpose of an SMS. They can explain the contents of an SMS and the processes of its implementation. Hale's SMS is a comprehensive and well-structured system, which makes it suitable for a comparison with other SMSs. This model provides a tool for assessing the completeness of an SMS. The performance of a safety management system can be derived from three groups of indicators: the initial risk based on incident or accident scenarios, the risk after insertion of safety barriers, and the delivery management for the barriers and controls. The three groups of indicators are not only present in Hale's SMS, but also correspond to the three groups of SMS models.

 **Words and Expressions**

 Unit Eighteen

physiology [ˌfɪziˈɒlədʒi] *n.* 生理学；生理机能
hygiene [ˈhaɪdʒiːn] *n.* 卫生；卫生学；保健法
the National Safety Council 国家安全委员
biological [ˌbaɪəˈlɒdʒɪk(ə)l] *adj.* 生物的；生物学的
habitability [hæbɪtəˈbɪlɪti] *n.* 可居住；适于居住
ergonomics [ˌɜːɡəˈnɒmɪks] *n.* 工效学；人类工程学
biomechanics [ˌbaɪəʊməˈkænɪks] *n.* 生物力学；生物机械学
construction site 建筑工地
oil drilling rig 石油钻井平台
guesswork [ˈɡeswɜːk] *n.* 猜测；臆测
mythology [mɪˈθɒlədʒi] *n.* 神话；神话学
superstition [ˌsuːpəˈstɪʃ(ə)n] *n.* 迷信；迷信行为
recommendation [ˌrekəmenˈdeɪʃn] *n.* 推荐；建议

systematic framework 系统框架
management [ˈmænɪdʒmənt] *n.* 管理；操纵
organizational [ˌɔːɡənaɪˈzeɪʃənl] *adj.* 组织的；编制的
constituent [kənˈstɪtʃuənt] *n.* 成分；选民；委托人；*adj.* 构成的；选举的
continuous improvement 持续改进
imperative [ɪmˈperətɪv] *adj.* 必要的；不可避免的；*n.* 必要的事；命令
mechanism [ˈmekənɪzəm] *n.* 机制；原理
industrial context 工业环境
accident analysis 事故分析
regulation [ˌreɡjuˈleɪʃn] *n.* 管理；规则；校准；*adj.* 规定的；平常的
consequence [ˈkɒnsɪkwəns] *n.* 结果；推论
awareness [əˈweənəs] *n.* 意识；认识；*n.* 人群对品牌或产品的认知
theoretical modelling 理论模型
event sequence 事件序列
compliance management 合规性管理

参考文献

[1] 朱晋科，周本友 . 矿业英语 [M]. 北京：中国大学出版社，1989: 263–321.

[2] 卢卫永，付玲毓，蒋国安 . 采矿工程英语 [M]. 北京：中国矿业大学出版社，2022：157–158, 162–165, 169–170.

[3] 人事部人事考试中心，国家外国专家局培训中心 . 职称英语（理工类）[M]. 沈阳：辽宁大学出版社，2005: 174–175.

[4] 唐敏康，邓晓宇 . 安全科学及工程专业英语 [M]. 北京：冶金工业出版社，2011：21–25, 141–143.

[5] 贾进章 . 安全工程专业英语 [M]. 北京：煤炭工业出版社，2014：30–33, 54–58.

[6] 司鹄 . 安全工程专业英语 [M]. 北京：机械工业出版社，2007：44–46, 117–122, 175–180.

[7] 樊晓运 . 安全工程专业英语 [M]. 北京：化学工业出版社，2012: 23–27, 118–123.

[8] 邓奇根，高建良，刘明举 . 安全系统工程（双语）[M]. 徐州：中国矿业大学出版社，2011：3–14, 71–81.

[9] 黄志安，张英华 . 安全工程专业英语 [M]. 北京：机械工业出版社，2018: 111–114, 187–192.

[10] 李霞，郭凤霞，王振会等 . 安全工程专业英语读物 [M]. 北京：气象出版社，2021: 1–3, 44–46.